Lecture Notes
in Business Information Processing   **244**

More information about this series at http://www.springer.com/series/7911

Paolo Ceravolo · Stefanie Rinderle-Ma (Eds.)

Data-Driven Process Discovery and Analysis

5th IFIP WG 2.6 International Symposium, SIMPDA 2015
Vienna, Austria, December 9–11, 2015
Revised Selected Papers

 Springer

Editors
Paolo Ceravolo
Università degli Studi di Milano
Crema
Italy

Stefanie Rinderle-Ma
Research Group Workflow Systems and
 Technology
Universität Wien
Vienna
Austria

ISSN 1865-1348   ISSN 1865-1356   (electronic)
Lecture Notes in Business Information Processing
ISBN 978-3-319-53434-3   ISBN 978-3-319-53435-0   (eBook)
DOI 10.1007/978-3-319-53435-0

Library of Congress Control Number: 2017930618

Printed on acid-free paper

This Springer imprint is published by Springer Nature
The registered company is Springer International Publishing AG
The registered company address is: Gewerbestrasse 11, 6330 Cham, Switzerland

Preface

The rapid growth of organizational and business processes managed via information systems has made available a big variety of data that consequently created a high demand for making data analysis techniques more effective and valuable. The fifth edition of the International Symposium on Data-driven Process Discovery and Analysis (SIMPDA 2015) was conceived to offer a forum where researchers from different communities and the industry can share their insights in this hot new field. As a symposium, SIMPDA fosters exchanges among academic researchers, industry, and a wider audience interested in process discovery and analysis. The event is organized by the IFIP WG 2.6. This year the symposium was held in Vienna.

Submissions cover theoretical issues related to process representation, discovery, and analysis or provide practical and operational experiences in process discovery and analysis. To improve the quality of the contributions, the symposium fostered discussions during the presentation. Papers are pre-circulated to the authors, who are expected to read them and make ready comments and suggestions. After the event, authors have the opportunity to improve their work extending the presented results. For this reason, authors of accepted papers and keynote speakers were invited to submit extended articles to this post-symposium volume of LNBIP. There were 22 submissions and eight papers were accepted for publication.

During this edition, the presentations and the discussions frequently focused on the adoption of process mining algorithms in conjunction and coordination with other techniques and methodologies. The current selection of papers underlines the most relevant challenges that were identified and proposes novel solutions and approaches facing these challenges.

In the first paper, "A Framework for Safety-Critical Process Management in Engineering Projects," Saimir Bala et al. present a framework for process management in complex engineering projects that are subject to a large amount of constraints and make use of heterogeneous data sources that must be monitored consistently to indicators and business goals.

The second paper, by Alfredo Bolt et al., is titled "Business Process Reporting Using Process Mining, Analytic Workflows and Process Cubes: A Case Study in Education." It illustrates in detail a case study where state-of-the-art process-mining techniques are used to periodically produce automated reports that relate the actual performance of students of a Dutch university to their studying behavior. Based on two evaluations, the authors discuss the acceptance level and the quality achieved by reports generated using process mining tools.

The third paper, by Bart Hompes et al., is titled "Detecting Changes in Process Behavior Using Comparative Case Clustering" and presents a novel comparative case-clustering approach that is able to expose changes in behavior. Valuable insights can be gained and process improvements can be made by finding those points in time where behavior changed and the reasons why this happened.

The fourth paper by Parabhakar Dixit et al., "Using Domain Knowledge to Enhance Process Mining Results," proposes a verification algorithm to verify the presence of certain constraints in a process model. This is particularly relevant when the user has certain domain expertise that should be exploited to create better process models. The outcome of the proposed approach is a Pareto front of process models based on the constraints specified by the domain expert and by common quality dimensions of process mining.

The fifth paper by Stefan Bunk et al., "Aligning Process Model Terminology with Hypernym Relations," faces the challenge of using a consistent terminology to label the activities of process models. To support this task, the authors defined two techniques to detect specific terminology defects, namely, process hierarchy defects and object hierarchy defects, and give recommendations to align them with hypernym hierarchies.

The sixth paper by Andreas Solt et al., "Time Series Petri Net Models: Enrichment and Prediction," aims at conciliating time series and Petri net models to provide operational support to predict the performance of individual cases and the overall business process considering seasonal effects.

The seventh paper by Theresia Gschwandtner, "Visual Analytics Meets Process Mining: Challenges and Opportunities," shows that the combination of visual data exploration with process mining algorithms makes complex information structures more comprehensible and facilitates new insights. Thus, before and during the application of automated analysis methods, such as process mining algorithms, the analyst needs to investigate how visual analytics can integrate a knowledge discovery environment.

The eighth paper by Thomas Vogelgesang and Jurgen Appelrath, "A Relational Data Warehouse for Multidimensional Process Mining," studies how to implement a relational database supporting OLAP operations for process mining.

We gratefully acknowledge the strong research community that gathered around the research problems related to process data analysis and the high quality of their research work, which is hopefully reflected in the papers of this volume. We would also like to express our deep appreciation of the referees' hard work and dedication. Above all, thanks are due to the authors for submitting the best results of their work to the Symposium on Data-driven Process Discovery and Analysis.

We are very grateful to the Università degli Studi di Milano and to IFIP for their financial support, and to the University of Vienna and the Austrian Computer Society for hosting the event.

January 2017

Paolo Ceravolo
Stefanie Rinderle-Ma

Organization

Chairs

Paolo Ceravolo   Università degli Studi di Milano, Italy
Stefanie Rinderle-Ma   Universität Wien, Austria

Advisory Board

Ernesto Damiani   Università degli Studi di Milano, Italy
Erich Neuhold   University of Vienna, Austria
Maurice van Keulen   University of Twente, The Netherlands
Philippe Cudré-Mauroux   University of Fribourg, Switzerland
Marcello Leida   Ebtic (Etisalat Bt Innovation Centre), UAE

SIMPDA Award Committee

Gregorio Piccoli   Zucchetti Spa, Italy
Paolo Ceravolo   Università Degli Studi Di Milano, Italy
Maria Leitner   Austrian Institute of Technology, Austria
Marcello Leida   Ebtic (Etisalat Bt Innovation Centre), UAE

Web and Publicity Chair

Fulvio Frati   Università degli Studi Di Milano, Italy

Program Committee

Mohamed Achemlal   University of Bordeaux, France
Marco Anisetti   Università degli Studi di Milano, Italy
Irene Vanderfeesten   Eindhoven University of Technology, The Netherlands
Claudio Ardagna   Università degli Studi di Milano, Italy
Helen Balinsky   Hewlett-Packard Laboratories, UK
Mirco Bianco   Metroconsult Roberto Dini and Partners, Italy
Joos Buijs   Eindhoven University of Technology, The Netherlands
Antonio Caforio   Università del Salento, Italy
Carolina Chiao   University of Ulm, Germany
Tony Clark   Middlesex University, UK
Barabara Weber   University of Innsbruck, Austria
Paul Cotofrei   University of Neuchâtel, Switzerland
Philippe Cudre-Mauroux   University of Fribourg, Switzerland

Contents

A Framework for Safety-Critical Process Management in Engineering Projects

Saimir Bala[1], Cristina Cabanillas[1], Alois Haselböck[2], Giray Havur[1(✉)],
Jan Mendling[1], Axel Polleres[1], Simon Sperl[2], and Simon Steyskal[1,2]

[1] Vienna University of Economics and Business, Vienna, Austria
{saimir.bala,cristina.cabanillas,giray.havur,jan.mendling,
axel.polleres}@wu.ac.at
[2] Siemens AG Österreich, Corporate Technology, Vienna, Austria
{alois.haselbock,simon.steyskal}@siemens.com

Abstract. Complex technical systems, industrial systems or infrastructure systems are rich of customizable features and raise high demands on quality and safety-critical aspects. To create complete, valid and reliable planning and customization process data for a product deployment, an overarching engineering process is crucial for the successful completion of a project. In this paper, we introduce a framework for process management in complex engineering projects which are subject to a large amount of constraints and make use of heterogeneous data sources. In addition, we propose solutions for the framework components and describe a proof-of-concept implementation of the framework as an extension of a well-known BPMS.

Keywords: Adaptation · Compliance · Engineering · Process Management · Resource management · Unstructured data

1 Introduction

Deployments of technical infrastructure products are a crucial part in the value-creation chain of production systems for large-scale infrastructure providers. Examples of a large-scale, complex engineering process in a distributed and heterogeneous environment are the construction of a railway system comprising electronic interlocking systems, European train control systems, operator terminals, railroad crossing systems, etc.; all of the systems are available in different versions using a variety of technologies. It is often necessary to offer, customize, integrate, and deliver a subset of these components for a particular customer project, e.g. the equipment of several train stations with an electronic switching

---

This work is funded by the Austrian Research Promotion Agency (FFG) under grant 845638 (SHAPE): http://ai.wu.ac.at/shape-project/.

P. Ceravolo and S. Rinderle-Ma (Eds.): SIMPDA 2015, LNBIP 244, pp. 1–27, 2017.
DOI: 10.1007/978-3-319-53435-0_1

unit in combination with a train control system based on radio communication for a local railway company. Configurators are engineering tools for planning and customizing a product. Each subsystem comes with its own, specialized engineering and verification tools. Therefore, configuring and combining these subsystem data to a coherent and consistent system has to follow a complex, collaborative process.

A challenge is the management and monitoring of such complex, yet mostly informally described, engineering processes that involve loosely integrated components, configurators and software systems [1]. Nowadays, many of the steps required (e.g. resource scheduling, document generation, compliance checking) tend to be done manually, which is error-prone and leads to high process execution times, hence potentially affecting costs in a negative way.

In this paper, we explore this domain and present a framework for process management in complex engineering processes that includes the formalization of human-centric process models, the integration of heterogeneous data sources, rule enforcement and compliance checking automation, and adaptability, among others. The framework has been defined from an industry scenario from the railway automation domain. Furthermore, we describe solutions to support the functionalities required by every framework component as well as a proof-of-concept implementation of the framework that can be integrated with an existing Business Process Management System (BPMS). The goal is to help to develop ICT support for more rigorous and verifiable process management.

The rest of the paper is organized as follows. Section 2 delves into the problem and derives system requirements. Section 3 presents the framework and solutions for its components. Section 4 describes a proof-of-concept implementation. Section 5 summarizes related work. Finally, Sect. 6 draws conclusions from this work and outlines ideas for future extensions.

2 Motivation

In the following, we describe an industry scenario that exemplifies the characteristics of complex engineering processes and define a set of system requirements for the challenges identified in it.

2.1 Industry Scenario

Activities to create complete, valid and reliable planning, and customization process data for a product deployment are part of an overarching engineering process that is of crucial importance for the success of a project in a distributed, heterogeneous environment. Figure 1 depicts a generic engineering process for building a new infrastructure system in the railway automation domain modeled with Business Process Model and Notation (BPMN) [2].

The engineering process itself is represented in the pool *Railway automation unit* and comprises the building and testing of the system. The pool *Resource planning unit* as well as the activities depicted in gray represent a meta-process

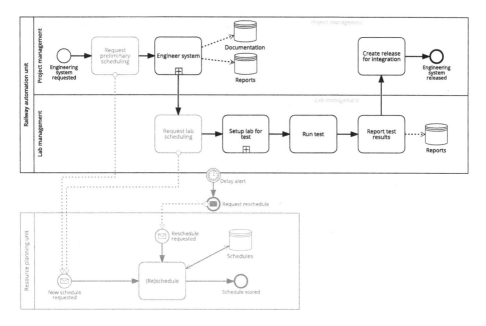

Fig. 1. Generic engineering process in the railway automation domain

comprising scheduling activities that are performed in the background in order to enable the completion of the engineering process in compliance with a set of restrictions (temporal and logistics, among others) while making an appropriate use of the resources available. Resource allocation is of great importance to large-scale engineering processes in which a large variety of different resources, ranging from laboratories and specific hardware to engineers responsible for the correct execution of the process, are involved and unexpected situations may have critical consequences (e.g., delays resulting in unplanned higher costs).

Hence, the first step consists of scheduling the building of the system. Building the system is, in turn, a process composed of several activities (potentially operating on different levels of abstraction) each involving a large variety of different resources, data sources, and data formats used. Specifically, the customer provides input data in form of, e.g., XML documents representing railway topology plans, signal and route tables, etc., which are used by the engineers to configure the product. Typically, several configuration tools are involved in that process too, complemented by version control and documentation activities. The result is a set of data of various kinds and formats (i.e., XML, JSON, and alike) such as bill of material (BOM), assembly plans, software configuration parameters, and all other documents and information required for the testing, integration, and installation of the system. Additionally, we map all gathered data to a common extendable RDF model in order to make use of standard data integration and processing strategies from the Semantic Web (e.g., OWL, SPARQL, SHACL, etc.). The engineering project manager orchestrates and monitors these engineering tasks. Besides, further data is generated during the execution of the subprocess *Engineer system* in the form of, e.g., emails exchanged between the process participants.

Once the system is built, it must be tested before it is released for its use. That procedure takes place in laboratories and comprises two phases: the test setup and run phases. Like before, it is necessary to schedule these activities taking into consideration the setting and all the restrictions for the execution of the activities. The setting is the following: there are several space units distributed into two laboratories and several units of different types of hardware for conducting the testing. The employees of the organization involved in these activities are specialized in the execution of specific testing phases for specific types of systems, i.e. there may be an engineer who can perform the setup for a system S_1 and the test execution for a system S_2, another engineer who cannot be involved in the testing of system S_1 but can perform the two activities for the system S_2, and so on. As for the restrictions, they are as follows: each task involved in these two phases requires a specific set of resources for its completion. In particular, the setup tasks usually require one employee working on one unit of a specific type of hardware in a laboratory, and the run activity usually requires several employees for its execution. Besides, a test can only be executed if the whole setup takes place in the same laboratory. In addition, for the scheduling it is necessary to take into account that other instances of the same or different processes might be under execution at the same time and they might share resources.

The setup and the run test activities will then be executed according to the plan. Similar to the engineering step, data comprising the results of the tests, emails, Version Control System (VCS) file updates and the like, is generated during the testing steps. Railway projects also generate other types of data which play a role in the running system, e.g., cut plans, signals form the tracks, actual user data of the system, etc. The latter can be useful when it comes to monitoring safety-critical processes during their execution. Given the great number of software tools involved in the process, our scenario focuses more on the software engineering aspect of the railway domain. Hence, we use VCS logs to track the evolution of the artifacts that are produced during such software engineering process. Nevertheless, it can be extended towards all kinds of artifacts that are stored in VCSs, such as the different versions of outputs from engineering tools.

When the testing of the system is finished, a final report is written and archived with the information generated containing the description of the test cases, test data, test results, and the outline of the findings. Responsible for the final version of this report is the testing project manager. Finally, the engineering project manager deploys a complete and tested version of the engineering system and the integration team takes over the installation of the product.

Note that unexpected situations may cause delays in the completion of any of the activities involved in the engineering process. It is important to detect such delays as soon as possible in order to properly schedule the use of resources and figure out when the process can be finished under the new circumstances. Therefore, rescheduling may be required at any point, involving all the aforementioned restrictions and possibly new ones.

2.2   Challenges and Technical Requirements

A number of issues are involved in the industry scenario previously described when it comes to automating its execution. From the analysis of the process description, the following challenges have been identified:

Challenge 1: Integrated description of processes, constraints, resources and data. Operating with processes like the one described before implies taking not only the order of execution of the process activities and behavioural constraints typically enforced in the process model into consideration, but also information related to other business processes perspectives, such as resources and data requirements, as well as regulations affecting, e.g., the use of these. Several formal languages are at hand for describing processes [2], constraints [3], resources [4] and data (e.g. XML) separately. However, a challenge is to define all them in an integrated manner with a model that provides rich querying capabilities to support analysis automation, status monitoring or respectively, the verification of constraints and consistency.

Therefore, a system for automating processes like the engineering process would require *an integrated semantic model to describe and monitor processes, resources, constraints and data (RQ1).*

Challenge 2: Integration and monitoring of structured and unstructured data. To a high degree, engineering steps are the input for state changes of the process, often only visible as manipulation of data. Hence, a engineering process must also incorporate these data smoothly for monitoring control flow, version updates, data storage, and email notification. To this end, various types of systems have to be integrated including their structured (e.g., logs from tools, or databases) and unstructured data (e.g., by mail traffic, or ticketing systems). Up until now, these data are hardly integrated and are monitored mostly manually.

Therefore, techniques to gather relevant information from unstructured or semi-structured data sources, such as emails, VCS repositories and data from tools, and transform them into understandable data structures must be put in place, i.e. a system for automating processes like the engineering process would require *mechanisms to detect and extract structured from unstructured process data (RQ2).*

Challenge 3: Documentation of safety-critical, human and data aspects and compliance checking. Engineering projects have time-critical phases typically prone to sloppy documentation and reduced quality of results. Many of the process steps are required to be documented in prescribed ways by standards and regulations (e.g. SIL [5]). Considerable amount of time is spent in the manual documentation of process steps as well as in the integration of the documentation of separate modules for generating final reports. Furthermore, in such safety-critical environments the use of resources must be optimized and rules must be enforced and their fulfillment ensured. For instance, in our industry scenario we can observe a typical series of data management steps, including: check in a new version of a data file, inform the subsequent data engineer, confirm and document this step in the process engine, etc. The latter two steps could be done automatically once the process engine has detected the check-in into the VCS. This automation would also lead to a significant decrease of the overall execution time as well as a potential reduction in the number errors typically caused by human mistakes.

Therefore, a system for automating processes like the engineering process would require *a method for flexible document generation (RQ3)* as well as *reasoning mechanisms (RQ4)* and *monitoring capabilities (RQ5)* for automating resource allocation and compliance checking.

Challenge 4: Be ready for changes. Despite engineering process definitions are quite stable and might remain unchanged for a long time, an automatic monitoring and a thorough analysis of process models and executions may lead to the discovery of potential improvement points to make processes simpler and less error-prone. Similarly, changes in the schedule of activities and resources can be necessary at any time due to a number of reasons including delays or unexpected unavailability of resources, among others. Currently, all these adaptations require manual work and are prone to errors.

Consequently, methods to detect and deal with changes must be put in place. This might include the monitoring of process executions and the analysis of the generated execution data (e.g. with process mining techniques) to anticipate delays as well as the need of having available mechanisms capable of reallocating the resources according to changeable requirements and circumstances. Therefore, besides requirements RQ4 and RQ5, a system for automating processes like the engineering process would also require *adaptation procedures to react to changing conditions (RQ6)*.

Challenge 5: Acceptance and human factors. The overall process management needs to be set up in a non-obtrusive way, such that engineers executing the processes find it useful and easy to use. This is a specific challenge in safety-critical systems, which are developed with a tight timeline. It calls for a design that integrates existing tools and working styles instead of introducing new systems.

Therefore, the automation of processes like the engineering process would require *an integrated system (RQ7)* that provides all the functionality, which involves general features of a BPMS (e.g. process modeling and execution) extended to support the demands of safety-critical, human- and data-centric processes as described in our industry scenario.

3   Framework

We have designed a framework that provides the support required to address the challenges identified in complex engineering processes. It consists of a data model and five functional modules that interact with a BPMS, as depicted in Fig. 2 using the Fundamental Modeling Concepts (FMC) notation[1].

---

[1] http://www.fmc-modeling.org/.

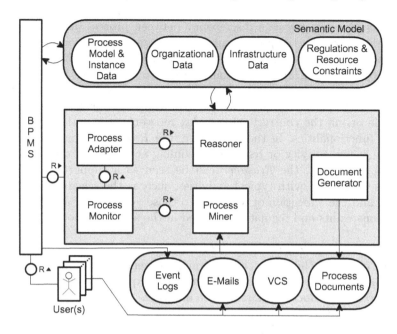

Fig. 2. Proposed framework for process management in complex engineering projects

In order to support *RQ1*, a semantic model encompassing the various types of domain data that must be represented and manipulated must be defined. Hence, this model stores all static and dynamic data used by the BPMS and by the functional components. Specifically, these data include: process models and their instances, organizational data related to human resources, infrastructure data related to non-human resources, and constraints derived from regulations and norms (e.g. SIL) as well as further requirements related to the utilization of resources. The semantic model implicitly operates as a communication channel between the BPMS and all the functional modules and hence, all of them must have read & write access to the model.

Typical functionality of a BPMS include modeling and executing processes. Information about process instances is usually stored in event logs generally including, among others, temporal and resource information related to the execution of the process activities [6]. In addition to that structured information, as described in Sect. 2.1, several kinds of unstructured and semistructured data are generated during the execution of complex engineering processes, e.g. emails, VCS files and reports. All the data produced during process execution must be analyzed in order to detect anomalies (e.g., deviations from the expected behavior).

The *Process Miner* component of our framework tries to discover as much data relevant to the current state of a process execution as possible, performs the transformations required as specified by *RQ2*, and communicates the information extracted to the *Process Monitor* (*RQ5*) periodically under request. In case the *Process Monitor* reveals a discrepancy between process instance data and the data discovered by the process miner (e.g., a delay), it informs the *Process Adapter* about the discrepancy. The *Process Adapter* analyzes the deviation and responds by proposing an adaptation solution to the BPMS in order to put the process back into a coherent and consistent state, as specified in *RQ6*. The adaptation may consist of small changes that can be performed directly on the BPMS side or, on the contrary, of complex recovery actions that may require reasoning functionalities. In the latter case, the *Reasoner* comes into play by, e.g., doing a new activity or resource scheduling according to the new domain conditions. Therefore, the *Reasoner* can be seen as a supportive component that helps the BPMS with typical activities, such as the scheduling of process activities, and the allocation of resources to those activities in accordance with resource constraints and regulations defined in the semantic model. This covers *RQ4*.

Finally, the *Document Generator* of the framework provides support for *RQ3* by helping to fill out the documents that must be generated as output of process activities. As mentioned before, this automation is expected to decrease reporting errors, especially in documents related to auditing.

The design of the framework as an extension of the functionality present in existing BPMSs attends to *RQ7* and hence, it intends to increase the acceptance by users familiar with Business Process Management (BPM).

In the following, we describe our solution for the implementation of the functionality provided by the most domain specific components of the framework.

3.1   Semantic Model

Aiming at automation, we believe that Semantic Web technologies provide the most appropriate means for (i) integrating and representing domain-specific (heterogeneous) knowledge in a consistent and coherent format, and (ii) querying and processing integrated knowledge.

Therefore, following the METHONTOLOGY approach [7], we have developed an engineering domain ontology [8] that integrated three different domains of interest relevant for our approach, namely: (i) engineering domain and organizational (i.e. resource-related) knowledge; (ii) business processes; and (iii) regulations and policies [9].

Representing Infrastructural and Organizational Knowledge. One of the first steps for developing an ontology according to the METHONTOLOGY approach involves the definition of an *Ontology Requirements Specification Document* [10]. In order to address the requirements gathered throughout that process we decided to adopt parts of the organizational meta model described in [11] and enriched it with concepts for modeling teams [12] (cf. Fig. 3) for representing infrastructural & organizational knowledge. Using these two meta models presents an advantage. Specifically, the organizational meta model described in [11] has been used to design a language for defining resource assignment conditions in process models called Resource Assignment Language (RAL) [4]. As was shown in [13], that language can be seamlessly integrated in existing process modeling notations, such as BPMN, thus enriching the process models with expressive resource assignments that cover a variety of needs described by the creation patterns of the well-known workflow resource patterns [11]. Furthermore, a graphical notation was later designed with the same expressive power as RAL in order to help the modeler to define resource assignments in process models [14]. The meta model for teamwork assignment was also considered to develop an extension of RAL called RALTeam [12], which, however, lacks a graphical notation so far. Therefore, if support for these expressive notations were introduced in the BPMS, the ontology would support them at the same time as it supports less expressive means of assigning resources to process activities (e.g. based on organizational positions).

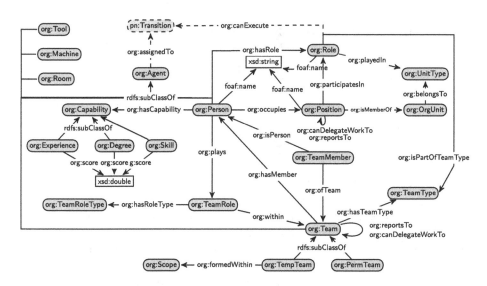

Fig. 3. Ontology for infrastructural & organizational knowledge.

Representing Business Processes. Driven by the requirements of our resource allocation approach (cf. Sect. 3.2), we decided to transform BPMN models into timed Petri nets [15] as an intermediary format for reasoning tasks (e.g., scheduling [16]) by using the transformation proposed in [17,18], and store these Petri nets in our ontology. Note that the user only interacts with the BPMN model while using the system as we use the timed Petri net representation internally. There are several reasons for using Petri nets for process modeling [19], namely:

– *Clear and precise definition:* Semantics of the classical Petri net is defined formally.

- *Expressiveness:* The primitives needed to model a business process (e.g. routing constructs, choices, etc.) are supported.
- *Tool-independent:* Petri nets have mappings to/from different business modelling standards [20]. Moreover, this immunizes our ontology from changes in business modeling standards.

For modeling Petri nets themselves we adopted selected concepts of the Petri Net Markup Language (PNML) [21] and represented them in terms of an RDFS ontology (cf. Fig. 4).

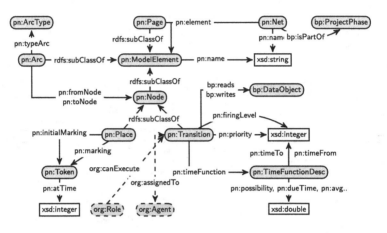

Fig. 4. Ontology for representing processes and process instances.

Extracting and Specifying Compliance Rules. One of the most important aspects of dealing with safety-critical human- and data-centric processes is providing means for proving that business processes comply with relevant guidelines such as domain-specific norms and regulations, or workflow patterns. As illustrated in Fig. 5 and described in [22], establishing proper compliance checking functionalities typically requires to extract and interpret a set of *Compliance Objectives* from respective *Compliance Requirements* first, before those objectives are specified in terms of *Compliance Rules/Constraints* using an appropriate specification language (i.e. a language capable of representing all types of compliance rules/constraints relevant for the respective domain of interest). Specified compliance rules and constraints are then subsequently used by a monitoring/compliance checking engine for verifying correct and valid execution of business processes w.r.t. previously defined rules.

1. *Identifying Compliance Objectives:* Organizations have to deal with an increasing number of norms and regulations that stem from various compliance sources, such normative laws and requirements. In the railway domain processes have to be compliant with specific European Norms (i.e. 50126, 50128, and 50129). For example, EN50126 defines guidelines for managing

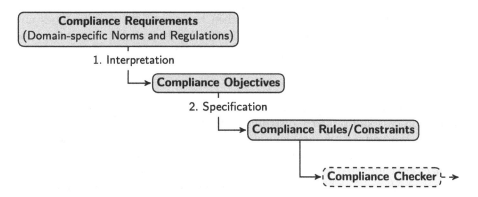

Fig. 5. General approach for business process compliance monitoring [22].

Reliability, Availability, Maintainability, and Safety (RAMS) of safety-critical business processes, where we extracted objectives, requirements, deliverables, and validation activities defined in all phases of the RAMS lifecycle [23,24].
2. *Representing Compliance Rules and Constraints:* Since all process relevant data are stored in RDF, we plan to utilize recent advancements in the area of constraint checking for RDF, i.e. the Shapes Constraint Language (SHACL) [25] for representing and validating identified compliance objectives. Since constraints expressed in SHACL are internally mapped to corresponding SPARQL queries, we can further complement our own compliance constraints with already existing approaches for compliance checking using SPARQL such as [26]. Compliance constraints expressed in SHACL are tightly integrated with the underlying ontology and can be validated during both design time and runtime[2].

3.2   Reasoner

The *reasoner* module supports our framework on top of the engineering domain ontology in two folds: (i) by performing automated resource allocation described in the declarative formalism Answer Set Programming (ASP) [27], and (ii) by querying the ontology for compliance checking. We also looked at other declarative programming paradigms (e.g., CLP(FD)), and our initial findings confirm the advantages of using ASP [28,29]. Some of these advantages are as follows:

– Compact, declarative and intuitive problem encoding
– Rapid prototyping and easy maintenance (e.g., no need to define heuristics)
– Complex reasoning modes (e.g., weight optimization)
– Ability to model effectively incomplete specifications
– Efficient solvers (e.g., *clingo*)

---

[2] For a more detailed introduction on utilizing SHACL for defining custom constraints, we refer the interested reader to [25].

In the literature, ASP is preferable when the size of the problem does not explode the grounding of the program [29,30]. We show that our resource allocation encoding in ASP is applicable to the problems of business processes at a real-world scale [16].

Resource Allocation. Resource allocation aims at scheduling activities of a business process and properly distributing available resources among scheduled activities. We address the problem of allocating the resources available to the activities in the running process instances in a time optimal way, i.e. process instances are completed in the minimum amount of time. Therefore, our resource allocation technique makes business process executions effective and efficient.

We encode the resource allocation problem in Answer Set Programming (ASP) [27], a declarative (logic programming style) paradigm. Its expressive representation language, efficient solvers, and ease of use facilitate implementation of combinatorial search and optimization problems (primarily *NP-hard*) such as resource allocation. Therefore, modifying, refining, and extending our resource allocation encoding is uncomplicated due to the strong declarative aspect of ASP. We use the ASP solver *clasp* [27] for our purpose as it has proved to be one of the most efficient implementations available [31]. Another complex reasoning extension supported in *clasp* are weight optimization statements [27] to indicate preferences between possible answer sets.

Resources are defined in the engineering domain ontology (the organizational data and the infrastructure data) where they are characterized by a *type* and can have one or more *attributes*. In particular, any resource type (e.g., org:Person in Fig. 3) is a subclass of org:Agent. The attributes are all of type rdf:Property. The organizational data consists of human resources, their attributes (e.g. their name, role(s), experience level, etc.) and current availabilities stored in the ontology. In the same fashion, infrastructure data represents material resources (i.e. tools, machines, rooms) and their availabilities. Resource allocation considers resources to be *discrete* and *cumulative*. Discrete resources are either fully available or fully busy/occupied. This applies to many types of resources, e.g. people, software or hardware. However, for certain types of infrastructure, availability can be partial at a specific point in time. For instance, a room's occupancy changes over time. Such a cumulative resource is hence characterized by its *dynamic* attribute (available space in the room) and it can be allocated to more than one activity at a time. Any statement in our ontology can be easily incorporated as the input of our problem encoding [32]. The following example shows an excerpt of organizational data in the ontology and its equivalent in ASP.

```
# Organizational data
:glen a org:Person; foaf:name"Glen";
      org:occupies testeng.
:testeng a org:Position; foaf:name "Test Engineer";
        org:participatesIn labmng.
:labmng a org:Role; foaf:name "Lab Management".
```

```
% Equivalent ASP encoding
person(glen). name(glen,"Glen").
occupies(glen,testeng).
position(testeng). name(testeng,"Test Engineer").
participatesIn(testeng,labmng).
role(labmng). name(labmng,"Lab Management").
```

There are two main operations under resource allocation: *Allocation of resources* and *re-allocation of resources as adaptation.*

Allocation of resources deals with the assignment of resources and time intervals to the execution of process activities. It can be seen as a two-step definition of restrictions. First, the so-called *resource assignments* must be defined, i.e., the restrictions that determine which resources can be involved in the activities [4] according to their properties. The outcome of resource assignment is one or more *resource sets* with the set of resources that can be potentially allocated to an activity at run time. The second step assigns cardinality to the resource sets such that different settings can be described.

As mentioned in Sect. 3.1, there exist languages for assigning resource sets to process activities [4, 33–35]. However, cardinality is generally disregarded under the assumption that only one resource will be allocated to each process activity. This is a limitation of current BPMS, which we overcome in our proposed framework.

The main temporal aspect is determined by the expected duration of the activities. The duration can be predefined according to the type of activity or calculated from previous executions, usually taking the average duration as reference. This information can be included in the executable process model as a property of an activity (e.g. with BPMN [2]) or can be modelled externally. As for the variable activity durations depending of the resource allocation, three specificity levels can be distinguished:

- *Role-based duration*, i.e., a triple $(activity, role, duration)$ stating the (minimum/average) amount of time that it takes to the resources within a specific resource set (i.e., cardinality is disregarded) to execute instances of a certain activity.
- *Resource-based duration*, i.e., a triple $(activity, resource, duration)$ stating the (minimum/average) amount of time that it takes to a concrete resource to execute instances of a certain activity.
- *Aggregation-based duration*, i.e., a triple $(activity, group, duration)$ stating the (minimum/average) amount of time that it takes to a specific group to execute instances of a certain activity. In this paper, we use *group* to refer to a set of human resources that work together in the completion of a work item, i.e., cardinality is considered. Therefore, a *group* might be composed of resources from different roles which may not necessarily share a specific role-based duration. An aggregation function must be implemented in order to derive the most appropriate duration for an activity when a group is allocated to it. The definition of that function is up to the organization.

Given (i) a process model and its instance data; (ii) organizational and infrastructure data; (iii) resource requirements, i.e. the characteristics of the resources that are involved in each activity to be allocated (e.g. roles or skills); (iv) temporal requirements; and (v) regulations such as access-control constraints [4], i.e. *separation of duties (SoD)* and *binding of duties (BoD)*, the ASP solver finds an optimal allocation. The aforementioned functionalities and the entire associated ASP encoding are detailed in [36].

While executing the process instance, changes may be introduced to input used for allocation. For instance, organizational data may change in case of absence, regulations may be modified or simply execution of activities may delay. In some cases, such a change in the ontology directly affects a running process instance, and therefore, the process monitor informs the process adapter. The process adapter may decide that the allocation should be performed again. *Adaptive re-allocation* is a key functionality in this scenario and it is indispensable for safety-critical, human- and data- centric process management.

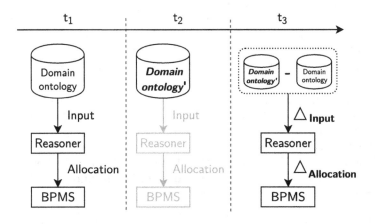

Fig. 6. Adaptive re-allocation timeline

Figure 6 shows this scenario in three consecutive time steps: After allocating resources to a process instance at t_1, some changes are introduced at t_2 that interfere with the original allocation, and hence, an adaptive reallocation is performed at t_3. The reasoner computes a delta allocation, i.e. the original allocation is preserved as much as possible. Therefore, some activities might be rescheduled, and others might be shifted and/or reallocated to some different resources in a *minimal* fashion.

Compliance Checking. As mentioned previously, we define compliance constraints over business processes using SHACL. In order to do so, we translate each compliance objective to a corresponding SPARQL query first, before embedding it in a respective constraint component, which itself can then be integrated into the ontology [24].

3.3   Process Monitor and Process Adapter

Changes and deviations to the processes may occur during execution. For instance, new rules and regulations may require the process to operate differently. We want our framework to be able to handle these unexpected events.

The process data and the evidence from the process miner are compared by the process monitor for detecting deviations. The main idea is that the process adapter is informed about the deviations, therefore it minimizes the impact of these deviations in the running process instances by offering a recovery strategy. The process monitor and process adapter address the requirements *RQ5* and *RQ6*, respectively. The process monitor is able to run several algorithms for monitoring both the process behaviour and the process compliance to rules and regulations. This is performed by checking the current process constraints against the data from our semantic model.

The process adapter is in charge of handling exceptions that arise from the process monitor. This component acts in two different ways: Either it *(i)* corrects process behaviour with minimal intervention, or, in case a more complex adaptation is required, *(ii)* it stops the process and notifies the reasoner for planning an adaptation.

3.4   Process Miner

Traditional process mining algorithms [6,37] are able to give valuable insights into the different perspectives of a business process. Process models inferred from log files can further be analyzed for bottlenecks, performance, deviations from the expected behaviour, compliance with rules and regulations, etc. Regardless of the perspective they aim at mining as well as the type of process modeling notation used to represent the outcome (declarative versus imperative process mining), all these process mining algorithms require properly structured data. Specifically, they must comply with the XES [38] meta model. Any of the existing process mining techniques is a candidate to be used for the implementation of the *Process Miner* component in regard to the functionality related to traditional process mining and the decision should be made according to the specific characteristics desired.

However, the biggest challenge of this component in our framework is to deal with unstructured and semistructured data generated in the execution of the process activities, generally in the form of VCS files and emails. Although it is hard to mine process models out of such unstructured or semistructured data, some approaches can be used to obtain valuable insights on them. Specifically, [39] allow for transforming semi-structured VCS logs to process activities which can be mined by classic mining algorithms. Poncin et al. [40] developed the FRASR framework, which is enables to answers engineering questions by applying process to software repositories. The challenge here is to identify the relevant events for the files, from a process mining point of view. Di Ciccio et al. [41] propose the MAILOFMINE approach to discover artful processes laying underneath email collections. Bala et al. [42] adopt a visualization approach by mining project Gantt charts from VCS logs.

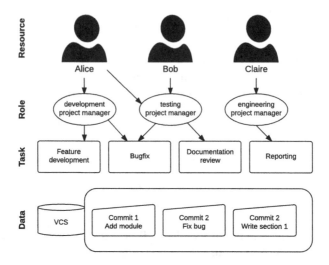

Fig. 7. Software project and resources

Driven by the fact that complex engineering process like the industry scenario described in Sect. 2.1 are resource-intensive, we have developed a novel approach to extract organizational roles from VCS repositories. VCSs have both structured and unstructured data. On the one hand, they explicitly provide information about the user who performed changes in some file(s) and the time at which she committed the new version(s). On the other hand, they have a textual part typically carrying a comment that explains the changes performed on the file(s). Note that these kind of data are similar to data from email. In fact, both emails and git comments have in common information about the user, the timestamp, and a textual description. Moreover, we use an ontology (cf. Sect. 3.1) to store mining insights. That is, we allow for the integration of all types of data that can be represented in RDF, including data coming from engineering tools. There are indeed tools in SVN [43] or Git [44] that allow for sending user commits as emails. Therefore, discovering roles out of such data (and especially when the outcome is combined with the result of mining activities) might help the *Process Monitor* to identify potential deviations regarding the resources that have actually performed specific tasks or manipulated certain information. Hence, it contributes to compliance checking.

Let us see an example of users who use VCS to collaborate on a software development project. Figure 7 shows a setting with three users named Alice, Bob and Claire. Alice is a development project manager. She works with her colleagues Bob and Claire. Alice is mainly responsible for *feature development*, but she is also involved in *testing* project-management team. Her tasks include the *development* of new features and fixing of related bugs. In her first *commit* she adds a new message where she describes her work. The message to describe her changes is "added new module to demo". In the first row of Table 1, identified by *commit id 1*, Alice's change is reported. Bob is part of the *testing* project-management team. His task is to ensure that the code submitted by the development team complies with existing standards and contains no errors. He discovers and fixes some minor bugs in Alice's code and informs Alice on further work needed on the analyzed features. Meanwhile, he commits his changes with *commit id 2* and comments "Modified the setup interface". Consequently, Alice reworks her code and commits a new version as reported in row 3 of Table 1, commenting her work with "Update application interface". As an *engineering project manager*, Claire takes over and starts to work on the documentation. She commits part of her work as in row 4. As the project continues, the work is accordingly stored in the log as shown in the table.

Table 1. Example of a VCS log

Id	User	Timestamp	Changed	Comment
1	Alice	2014-10-12 13:29:09	Demo.java rule.txt	Added new module to demo and updated rules
2	Bob	2014-11-01 18:16:52	Setup.exe	Modified the setup interface
3	Alice	2015-06-14 09:13:14	Demo.java	Update the application interface
4	Claire	2015-07-12 15:05:43	graph.svg todo.doc	Define initial process diagram & listed remaining tasks
...	...	...	...	...

Our approach leverages both on the file types and the comments of the users. We devise an algorithm that classifies users into a set of roles. For that purpose, we approach the role discovery problem as a classification problem. We define two methods: one based on user clustering and one based on commit clustering. A prior step for this is the feature selection. By looking at the commit data, we identify the following features:

- Total number of commits.
- Timeframe between the first and the last commit of a user (i.e. the time he has been working on the project).
- Commit frequency: total number of commits divided by the time frame.
- Commit message length: average number words in the commit comment.
- Keyword count: how often determinate words like "test" or "fix" are used
- Number of files changed.
- Affected file types: how often a file with a certain format (e.g., *.java, *.html) are modified by a user, relative to the total number of modified files

Then, we use the features for two machine learning algorithms. In the first approach we iterate through the users and cluster them using the k-means algorithm. Consequently we build classification models using decision trees. We then train three different datasets individually and cross validate the results. The second approach starts from the commits. The main idea here is that we do not want to assign users to a specific category. Rather, we allow for users having multiple roles and classify their contribution in each commit. We build user profiles that account for fractions of contributions of each user to the different classes. Classes used in the classification for the example described above would be: *Test, Development, Web, Backend, Maintenance, Refactor, Documentation, Design, Build, Data, Tool, Addition, Removal, vcsManagement, Automated, Merge*. Each commit is classified into one of the classes according to its features. Users who committed can be then classified by their commits. The classification can be done both manually and automatically: *(i)* rules can be manually inferred by looking for similarities between users with the same role; or *(ii)* an automated classification can be performed by using machine learning algorithms. For example, decision trees can be used for an automated classification. In this case the *commit* type percentages are used as features and the manually assigned roles as classes. As a further step, the resulting decision tree models from the different datasets can be cross-validated. The complete approach and its evaluation can be found in [45].

3.5   Document Generator

Safety-critical engineering systems require well-documented process steps. Engineers are in charge of clearly describing their tasks in such a way that it is possible to audit their work. These documents are often manually created. This has at least four drawbacks. First, their creation is laborious. Second, it is error-prone and misaligned in terms of language. Third, it is described at different levels of granularity. Fourth, it is difficult to process and audit afterwards.

A simple example is the following. Engineers need to work on a specific task and use predefined tools. Their tasks and their version tools are specified at the beginning of the project and must be consistent during the project's lifetime. Tool versions must usually be filled in the documentation generated, e.g. in reports. In a big engineering project, tools can be numerous and their versions are far from being user-friendly. This makes the risk of human mistakes very likely.

To assist engineering project managers in producing audible documentations, we have developed a customizable approach for partially automating document generation. In particular, it is able to fill in trivial information (e.g., tool version, user name, task to which the user is assigned, etc.) into word processor documents. Our document generation technique is based on templates. These templates consist of evolving documents and are automatically filled in during the workflow, and therefore enable flexible process verification. Our approach generates standard documents which are compatible with predefined word processor programs and can be opened and edited by them.

Document generation comprises four steps, depicted in Fig. 8 and explained next. A mapping function is first defined from a process activity to a document which is generated as its outcome. Afterwards, an interpreter is defined which is in charge of filtering the relevant process activities and variables. Process variables are used by the BPMS during the execution of the process. Examples of process variables can be the name of the user that is currently assigned to one activity, the name of the running process instance, and everything that adds data to the executing process in the BPMS. The interpreter is not strictly bound to a particular process nor to a particular template, and is defined externally. This supports changes both in the process and in the template. The writing in the document is triggered by a listener. A listener waits for activity events. As soon as an activity is submitted, the interpreter and the mapping function work together to generate *(variable key,value)* pairs in the document template. This is run iteratively on the document until all the trivial data is filled in.

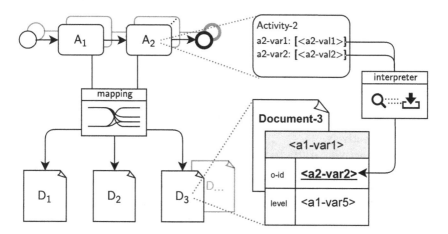

Fig. 8. Implementation showing how form values (process variables) from Camunda could look like in a generated document

4   Proof-of-concept Implementation

As a proof of concept we have implemented the main components of the framework discussed in Fig. 2. In this prototype, we aim to bring together functionality from reasoning, process mining and document generation. Figure 9 shows the software architecture that we use. It considers four main components which interoperate during the execution of a process activity.

Here, we describe the main components of the architecture and their interactions.

Camunda running process. We use the Camunda BPM engine as our BPMS. Camunda is an open source platform that allows for defining new components and for interacting with its APIs in a custom way. All the process instances that run into Camunda and their data are stored in log files. Camunda uses two main databases to store its logs: *(i)* a database for processes that are currently executing; and *(ii)* a database for historical information. These two databases can be queried through provided Java or REST APIs. Results are

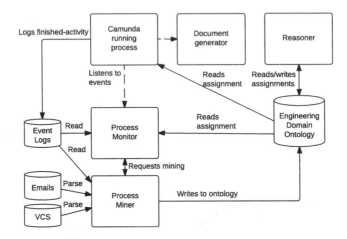

Fig. 9. Software architecture of the prototype

returned as either a set of Plain Old Java Objects(POJOs) or in the JSON format, respectively.

Before an activity starts to run, it first fetches the ontology which contains the set of assignments from existing resources to activities. Consecutively, a resource is assigned to the activity and thus can appear on their task list. When the resources complete their tasks, an event is triggered. This event is listened by the process miner and the document generator components, who can react accordingly. At the same time, the event is stored into the Camunda database of the running instances. Both the running processes database and the history database record similarly-structured data. Furthermore, they can be accessed using the same technology, i.e. the Camunda REST APIs. Hence, we abstract both these databases as a single database in Fig. 9 and denote it as *Event Logs*.

Reasoner. The reasoner module is implemented as a Java application connected to the Camunda process engine as an asynchronous service. We use Sesame, an open source framework for creating, parsing, storing, inferencing and querying over our ontology data. With respect to the request, the reasoner either performs resource allocation (cf. Fig. 10) by first translating the RDF data into the ASP language, solving the problem instance using the ASP solver *clasp*, and then writing the allocation results back to the triple store; or it validates all contained SHACL constraints and returns potential violation result back to the process engine.

Process Monitor. This component is in charge querying the status of the running processes in Camunda. In case a deviation occurs, for example, a process instance cannot be completed within the assigned schedule, the process monitor must signal out the anomaly. The process adaptation module can use this output to learn the status of the system and subsequently apply an adaptation. This component is implemented as a web client that can read

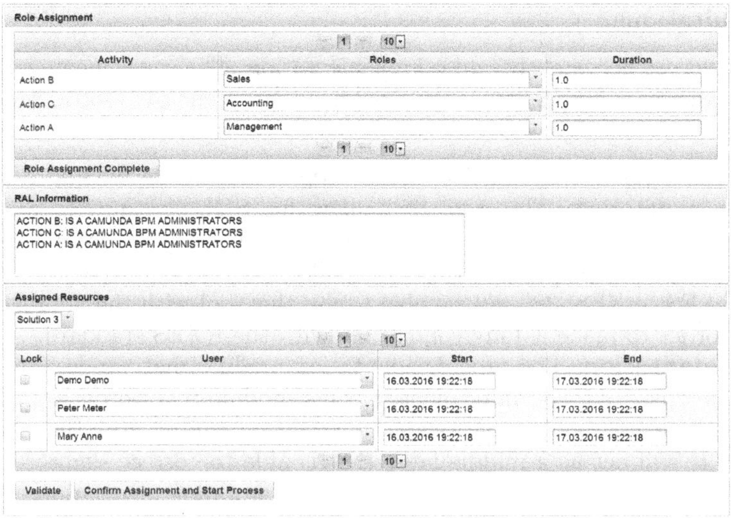

Fig. 10. Resource allocation interface

execution logs through the Camunda REST API. Results are returned in the JSON format which are then parsed into POJOs and can be processed by customized monitoring algorithms. In this case the communication happens through periodical queries to the database. An alternative to this is to implement an activity listener that notifies the process monitor whenever a task is completed.

Miner. The miner is in charge of running a number of mining algorithms on the logs from Camunda and from VCSs. Emails and commit messages can also be analysed by using the approaches discussed in [45]. This component is implemented as a web service, which can be called by the process monitor in order to understand how the activities being monitored have performed in the past. Mining algorithms can give new insights into the processes, like for instance actual execution times and several performance indicators of the process. This can contribute to the domain knowledge. Thus, they are stored again into the ontology as RDF.

Document generator. The document generator is in charge of listening to activity submissions and of collecting information from them with the final goal of creating textual documents. This component uses customizable event handlers to process changes of process variables and forms compiled by the users. It is implemented in Java and can be imported as a Java library into several other modules that require document generation from events.

4.1 Limitations

The architecture is currently under implementation. The components have been only individually tested. There is the need for a comprehensive software solution that integrates the single software components into one.

SQL console for querying Camunda logs. We are developing a tool for process monitoring. This tool will allow for SQL-like queries on top of Camunda logs. The approach involves mapping Camundas database schema to RXES [46]. In addition to this we are also developing a tool that can map from RXES to XES [38] and we plan to use this tool with the approach from [47] in order to make it fully compatible with the RXES standard.

Process adaptation. The process adaptation module that we describe in the framework is yet to be implemented. This module will be developed as an intermediate component between the process monitor and the Camunda engine. It will act as a middle layer that is able to correct slight deviations in the running process, without stopping the workflow. Deviations that are not adjustable may occur. In this case, this component will communicate the need for a schedule to the reasoner.

Connection to ontology. Our ontology is currently under improvement. We are planning to complete it with all the data from the engineering domain ontology (cf. Fig. 2). Furthermore, its connections to the various components are yet to be implemented.

User interfaces. We support for mining and monitoring techniques whose results are models that are generated out of data. User interfaces to visualize these data are required in order to easy the understanding of the mining results. Analogously, we plan to provide a fully fledged user interface for the reasoner component.

5   Related Work

The existing work on similar frameworks are from safety management [48–50], and decision support domains [51,52]. To best of our knowledge, there is no framework addressing all the seven requirements (cf. Sect. 2) that we identify. Therefore, we elaborate on the supporting literature.

Bowen and Stavridou [49] detail the standards concerned with the development of safety-critical systems, and the software in such systems. They identify the challenges of safety-critical computer systems, define the measures for the correctness of such systems and its relevance to several industrial application areas of, e.g. formal methods in railway systems, which is crucial for rigorous and coherent process management.

De Medeiros et al. [52] investigate the core building blocks necessary to enable semantic process mining techniques/tools. They conclude that semantic process mining techniques improve the conventional ones by using three elements, i.e., ontologies, model references from elements in logs/models to concepts in ontologies, and reasoners. Our framework supports such a high-level semantic analysis through our integrated semantic model and the reasoner module.

Wilke et al. [48] describe a framework for a holistic risk assessment in airport operations. They focus on coordination and cooperation of various actors through a process model derived in BPM, which helps determination of causal factors underlying operation failures, and detection and evaluation of unexpected changes. The holistic consideration of operations handling rules and regulations of their particular domain serves for ensuring compliance. Daramola et al. [50] describe the use of ontologies in a scenario requiring identification of security threats and recommendation of defence actions. Their approach not only help the quick discovery of hidden security threats but also recommend appropriate countermeasures via their semantic framework. By following this approach, they minimize the human effort and enable the formulation of requirements in a consistent way. In our framework we similarly monitor our ontology for compliance checking by querying the ontology via the queries derived from regulations.

Van der Aalst [53] introduced a Petri net based scheduling approach to show that the Petri net formalism can be used to model activities, resources and temporal constraints with non-cyclic processes. Several attempts have also been done to implement the problem as a constraint satisfaction problem. For instance, Senkul and Toroslu [54] developed an architecture to specify resource allocation constraints and a Constraint Programming (CP) approach to schedule a workflow according to the constraints defined for the tasks. Our framework addresses resource allocation via the reasoner module using ASP.

Zahoransky et al. [51] investigate operational resilience of process management. Their approach is proposed as a complementary approach to risk-aware BPM systems, which focuses on detecting the resilience properties of processes based on measures by mining process-logs for decision support to increase process resilience, and therefore provide flexibility. This approach enables agility in runtime and provides a solid foundation for process execution reliability. We address these aspects in our integrated system via the process miner and the process adapter.

6   Conclusions and Future Work

In this paper we have explored challenges of safety-critical human- and data-centric process management in engineering projects which are subject to a large amount of regulations and restrictions, i.e. temporal, resource-related and logistical restrictions, as described in the industry scenario. Our proposed framework addresses all the requirements derived from those challenges upon the general functionality of a BPMS, e.g. process adaptation, resource allocation, document generation and compliance checking.

This work is developed in cooperation with SIEMENS Austria who will be the primary user of the developed system. Our first proof of concept is implemented [55]. Next steps also involve putting in place adaptation mechanisms, implementing and integrating all the components into Camunda, and conducting a thorough evaluation of the implemented system w.r.t. real data from the railway domain.

References

1. Fleischanderl, G., Friedrich, G.E., Haselböck, A., Schreiner, H., Stumptner, M.: Configuring large systems using generative constraint satisfaction. IEEE Intell. Syst. **13**(4), 59–68 (1998)
2. OMG, "BPMN 2.0," recommendation, OMG (2011)
3. Governatori, G., Sadiq, S.: The journey to business process compliance. In: Cardoso, J., van der Aalst, W.M.P. (eds.) Handbook of Research on BPM, pp. 426–454. IGI Global, Hershey (2009)
4. Cabanillas, C., Resinas, M., del Río-Ortega, A., Ruiz-Cortés, A.: Specification and automated design-time analysis of the business process human resource perspective. Inf. Syst. **52**, 55–82 (2015)
5. Bozzano, M., Villafiorita, A.: Design and Safety Assessment of Critical Systems. CRC Press Taylor & Francis Group, Boca Raton (2010)
6. van der Aalst, W.: Process Mining: Discovery, Conformance and Enhancement of Business Processes. Springer, Heidelberg (2011)
7. Lopez, M.F., Perez, A.G., Juristo, N.: METHONTOLOGY: from ontological art towards ontological engineering. In: AAAI97 Symposium, pp. 33–40 (1997)
8. Cabanillas, C., Haselböck, A., Mendling, J., Polleres, A., Sperl, S., Steyskal, S.: Engineering Domain Ontology. SHAPE Project Deliverable (2016)
9. Steyskal, S., Polleres, A.: Defining expressive access policies for linked data using the ODRL ontology 2.0. In: SEMANTICS 2014, pp. 20–23 (2014)
10. Suárez-Figueroa, M.C., Gómez-Pérez, A., Villazón-Terrazas, B.: How to write and use the ontology requirements specification document. In: Meersman, R., Dillon, T., Herrero, P. (eds.) OTM 2009. LNCS, vol. 5871, pp. 966–982. Springer, Heidelberg (2009). doi:10.1007/978-3-642-05151-7_16
11. Russell, N., van der Aalst, W.M.P., ter Hofstede, A.H.M., Edmond, D.: Workflow resource patterns: identification, representation and tool support. In: Pastor, O., Falcão e Cunha, J. (eds.) CAiSE 2005. LNCS, vol. 3520, pp. 216–232. Springer, Heidelberg (2005). doi:10.1007/11431855_16
12. Cabanillas, C., Resinas, M., Mendling, J., Cortés, A.R.: Automated team selection and compliance checking in business processes. In: Proceedings of the International Conference on Software and System Process, ICSSP 2015, Tallinn, Estonia, pp. 42–51, 24–26 August 2015
13. Cabanillas, C., Resinas, M., Ruiz-Cortés, A.: RAL: a high-level user-oriented resource assignment language for business processes. In: Daniel, F., Barkaoui, K., Dustdar, S. (eds.) Business Process Management Workshops. Lecture Notes in Business Information Processing, vol. 99, pp. 50–61. Springer, Heidelberg (2011)
14. Cabanillas, C., Knuplesch, D., Resinas, M., Reichert, M., Mendling, J., Ruiz-Cortés, A.: RALph: a graphical notation for resource assignments in business processes. In: Zdravkovic, J., Kirikova, M., Johannesson, P. (eds.) CAiSE 2015. LNCS, vol. 9097, pp. 53–68. Springer, Heidelberg (2015). doi:10.1007/978-3-319-19069-3_4
15. Zuberek, W.: Timed petri nets definitions, properties, and applications. Microelectron. Reliab. **31**(4), 627–644 (1991)
16. Havur, G., Cabanillas, C., Mendling, J., Polleres, A.: Automated resource allocation in business processes with answer set programming. In: Reichert, M., Reijers, H.A. (eds.) Business Process Management Workshops. Lecture Notes in Business Information Processing, vol. 256, pp. 191–203. Springer, Heidelberg (2015)

17. Dijkman, R.M., Dumas, M., Ouyang, C.: Semantics and analysis of business process models in BPMN. Inf. Softw. Technol. **50**(12), 1281–1294 (2008)
18. Dijkman, R.M., Dumas, M., Ouyang, C.: "Formal semantics and analysis of BPMN process models using Petri nets," Technical report 7115, Queensland University of Technology (2007)
19. Van der Aalst, W.M.: The application of petri nets to workflow management. J. Circ. Syst. Comput. **8**(01), 21–66 (1998)
20. Lohmann, N., Verbeek, E., Dijkman, R.: Petri net transformations for business processes – a survey. In: Jensen, K., Aalst, W.M.P. (eds.) Transactions on Petri Nets and Other Models of Concurrency II. LNCS, vol. 5460, pp. 46–63. Springer, Heidelberg (2009). doi:10.1007/978-3-642-00899-3_3
21. Weber, M., Kindler, E.: The petri net markup language. In: Ehrig, H., Reisig, W., Rozenberg, G., Weber, H. (eds.) Petri Net Technology for Communication-Based Systems. LNCS, vol. 2472, pp. 124–144. Springer, Heidelberg (2003). doi:10.1007/978-3-540-40022-6_7
22. Ly, L.T., Maggi, F.M., Montali, M., Rinderle-Ma, S., van der Aalst, W.: Compliance monitoring in business processes: functionalities, application, and tool-support. Inf. Syst. **54**, 209–234 (2015)
23. Fuchsbauer, J.: "How to manage Processes according to the European Norm 50126 (EN 50126)." Bachelor thesis (2015)
24. Steyskal, S.: "Engineering Domain Ontology," project deliverable, Siemens (2016)
25. Knublauch, H., Ryman, A.: "Shapes Constraint Language (SHACL)," Working Draft (work in progress), W3C (2016). https://www.w3.org/TR/shacl/
26. Bouzidi, K.R., Faron-Zucker, C., Fies, B., Le Thanh, N.: An ontological approach for modeling technical standards for compliance checking. In: Rudolph, S., Gutierrez, C. (eds.) RR 2011. LNCS, vol. 6902, pp. 244–249. Springer, Heidelberg (2011). doi:10.1007/978-3-642-23580-1_19
27. Gebser, M., Kaminski, R., Kaufmann, B., Schaub, T.: Answer Set Solving in Practice. Morgan & Claypool Publishers, San Rafael (2012)
28. Brewka, G., Eiter, T., Truszczyński, M.: Answer set programming at a glance. Commun. ACM **54**(12), 92–103 (2011)
29. Dovier, A., Formisano, A., Pontelli, E.: A comparison of CLP(FD) and ASP solutions to NP-complete problems. In: Gabbrielli, M., Gupta, G. (eds.) ICLP 2005. LNCS, vol. 3668, pp. 67–82. Springer, Heidelberg (2005). doi:10.1007/11562931_8
30. Aschinger, M., Drescher, C., Friedrich, G., Gottlob, G., Jeavons, P., Ryabokon, A., Thorstensen, E.: Optimization methods for the partner units problem. In: Achterberg, T., Beck, J.C. (eds.) CPAIOR 2011. LNCS, vol. 6697, pp. 4–19. Springer, Heidelberg (2011). doi:10.1007/978-3-642-21311-3_4
31. Calimeri, F., Gebser, M., Maratea, M., Ricca, F.: Design and results of the fifth answer set programming competition. Artif. Intell. **231**, 151–181 (2016)
32. Eiter, T., Ianni, G., Krennwallner, T., Polleres, A.: Rules and ontologies for the semantic web. In: Baroglio, C., Bonatti, P.A., Małuszyński, J., Marchiori, M., Polleres, A., Schaffert, S. (eds.) Reasoning Web. LNCS, vol. 5224, pp. 1–53. Springer, Heidelberg (2008). doi:10.1007/978-3-540-85658-0_1
33. van der Aalst, W.M.P., ter Hofstede, A.H.M.: YAWL: yet another workflow language. Inf. Syst. **30**(4), 245–275 (2005)
34. Stroppi, L.J.R., Chiotti, O., Villarreal, P.D.: A BPMN 2.0 extension to define the resource perspective of business process models. In: CIbS 2011 (2011)
35. Cabanillas, C., Resinas, M., Mendling, J., Cortés, A.R.: Automated team selection and compliance checking in business processes. In: ICSSP, pp. 42–51 (2015)

36. Havur, G., Cabanillas, C., Mendling, J., Polleres, A., Haselböck, A.: Resource and data management service architecture. SHAPE Project Deliverable (2016)
37. Van Dongen, B.F., De Medeiros, A.K.A., Verbeek, H.M.W., Weijters, A.J.M.M., Van Der Aalst, W.M.P.: The ProM framework: a new era in process mining tool support. In: Ciardo, G., Darondeau, P. (eds.) ICATPN 2005. LNCS, vol. 3536, pp. 444–454. Springer, Heidelberg (2005). doi:10.1007/11494744_25
38. Verbeek, H.M.W., Buijs, J.C.A.M., Van Dongen, B.F., Van Der Aalst, W.M.P.: XES, XESame, and ProM 6. Information Systems Evolution. Lecture Notes in Business Information Processing, vol. 72, pp. 60–75. Springer, Heidelberg (2011)
39. Kindler, E., Rubin, V., Schäfer, W.: Activity mining for discovering software process models. Softw. Eng. **79**, 175–180 (2006)
40. Poncin, W., Serebrenik, A., Brand, M.V.D.: Process Mining Software Repositories. In: 15th European Conference on Software Maintenance and Reengineering, pp. 5–14 (2011)
41. Di Ciccio, C., Mecella, M., Scannapieco, M., Zardetto, D., Catarci, T.: MailOfMine - analyzing mail messages for mining artful collaborative processes. In: Aberer, K., Damiani, E., Dillon, T. (eds.) Data-Driven Process Discovery and Analysis. Lecture Notes in Business Information Processing, vol. 116, pp. 55–81. Springer, Heidelberg (2012)
42. Bala, S., Cabanillas, C., Mendling, J., Rogge-Solti, A., Polleres, A.: Mining project-oriented business processes. In: Motahari-Nezhad, H.R., Recker, J., Weidlich, M. (eds.) BPM 2015. LNCS, vol. 9253, pp. 425–440. Springer, Heidelberg (2015). doi:10.1007/978-3-319-23063-4_28
43. Pilato, C.M., Collins-Sussman, B., Fitzpatrick, B.W.: Version Control with Subversion. O'Reilly Media Inc, Sebastopol (2008)
44. Torvalds, L., Hamano, J.: Git: Fast version control system (2010). https://git-scm.com
45. Cabanillas, C., Bala, S., Mendling, J., Polleres, A.: "Combined method for mining and extracting processes, related events and compliance rules from unstructured data," Technical report, WU Vienna (2016)
46. van Dongen, B.F., Shabani, S.: "Relational XES: Data Management for Process Mining," BPM Cent. Rep. BPM-15-02 (2015)
47. Schönig, S., Cabanillas, C., Jablonski, S., Mendling, J.: Mining the organisational perspective in agile business processes. In: Gaaloul, K., Schmidt, R., Nurcan, S., Guerreiro, S., Ma, Q. (eds.) Enterprise, Business-Process and Information Systems Modeling. Lecture Notes in Business Information Processing, vol. 214. Springer, Heidelberg (2015)
48. Wilke, S., Majumdar, A., Ochieng, W.Y.: Airport surface operations: a holistic framework for operations modeling and risk management. Saf. Sci. **63**, 18–33 (2014)
49. Bowen, J., Stavridou, V.: Safety-critical systems, formal methods and standards. Softw. Eng. J. **8**(4), 189–209 (1993)
50. Daramola, O., Sindre, G., Moser, T.: A tool-based semantic framework for security requirements specification. J. UCS **19**(13), 1940–1962 (2013)
51. Zahoransky, R.M., Brenig, C., Koslowski, T.: Towards a process-centered resilience framework. In: 10th International Conference on Availability, Reliability and Security (ARES), pp. 266–273. IEEE (2015)
52. de Medeiros, A.K.A., Van der Aalst, W., Pedrinaci, C.: "Semantic process mining tools: core building blocks" (2008)
53. van der Aalst, W.: Petri net based scheduling. Oper. Res. Spektr. **18**(4), 219–229 (1996)

54. Senkul, P., Toroslu, I.H.: An architecture for workflow scheduling under resource allocation constraints. Inf. Syst. **30**, 399–422 (2005)
55. Bala, S., Havur, G., Sperl, S., Steyskal, S., Haselböck, A., Mendling, J., Polleres, A.: SHAPEworks: a BPMS extension for complex process management. In: BPM Demos (2016, to appear)

Business Process Reporting Using Process Mining, Analytic Workflows and Process Cubes: A Case Study in Education

Alfredo Bolt[1]([⊠]), Massimiliano de Leoni[1], Wil M.P. van der Aalst[1], and Pierre Gorissen[2]

[1] Eindhoven University of Technology, Eindhoven, The Netherlands
{a.bolt,m.d.leoni,w.m.p.v.d.aalst}@tue.nl
[2] Hogeschool van Arnhem en Nijmegen, Nijmegen, The Netherlands
pierre.gorissen@han.nl

Abstract. Business Process Intelligence (BPI) is an emerging topic that has gained popularity in the last decade. It is driven by the need for analysis techniques that allow businesses to understand and improve their processes. One of the most common applications of BPI is *reporting*, which consists on the structured generation of information (i.e., reports) from raw data. In this article, state-of-the-art process mining techniques are used to periodically produce automated reports that relate the actual performance of students of a Dutch University to their studying behavior. To avoid the tedious manual repetition of the same process mining procedure for each course, we have designed a *workflow* calling various process mining techniques using RapidProM. To ensure that the actual students' behavior is related to their actual performance (i.e., grades for courses), our analytic workflows approach leverages on *process cubes*, which enable the dataset to be *sliced* and *diced* based on courses and grades. The article discusses how the approach has been operationalized and what is the structure and concrete results of the reports that have been automatically generated. Two evaluations were performed with lecturers using the real reports. During the second evaluation round, the reports were restructured based on the feedback from the first evaluation round. Also, we analyzed an example report to show the range of insights that they provide.

Keywords: Business process reporting · Analytic workflows · Process mining · Process cubes · Education

1 Introduction

Business Process Reporting (BPR) refers to the provision of structured information about processes in a regular basis, and its purpose is to support decision makers. Reports can be used to analyze and compare processes from many perspectives (e.g., behavior, performance, costs, time). In order to be effective, BPR presents some challenges:

© IFIP International Federation for Information Processing 2017
Published by Springer International Publishing AG 2017. All Rights Reserved
P. Ceravolo and S. Rinderle-Ma (Eds.): SIMPDA 2015, LNBIP 244, pp. 28–53, 2017.
DOI: 10.1007/978-3-319-53435-0_2

1. It should provide insights using metric-based characteristics (e.g., bottlenecks, throughput time, resource utilization) and behavioral characteristics (e.g., deviations, frequent patterns) of processes.
2. It should be repeatable (i.e., not require great efforts to repeat the analysis).
3. It should be able to analyze the data with different granularity levels (i.e., analyze an organization as a whole or analyze its branches individually).

This paper shows through a case study how *process mining, analytic workflows* and *process cubes* can be concretely used for business process reporting, addressing the three challenges mentioned above.

The case study presented in this paper refers to a business-process reporting service at Eindhoven University of Technology. The service produces a report each quartile (i.e., two-month academic period) for each course that is provided with video lectures. The report is sent to the responsible lecturer and provides insights about the relations between the students' usage of video lectures and their final grades on the course, among other educational data analysis results.

Process mining is a relatively young research discipline that is concerned with discovering, monitoring and improving real processes by extracting knowledge from event logs readily available in today's systems [1]. Hundreds of different process mining techniques have been proposed in literature. These are not limited to process-model discovery and the checking of conformance. Also, other perspectives (e.g., data) and operational support (e.g., predictions) are included. Process mining is supported by commercial software (e.g., Disco[1], Celonis[2]) and academic software (e.g., ProM[3] [2]) tools.

Process mining allows the extraction of insights about the overall and inner behavior contained in any given process (e.g., a student taking a course). These insights can be collected and processed into reports. When thousands of different reports need to be produced (e.g., one for each course), it can be tedious and error-prone to manually repeat all the process-mining analyses to be incorporated in the reports. Analytic workflows can be used to fully automate analytic experiments such as the generation of an arbitrary number of reports. Process cubes can be used to scope and split the overall process data into the granularity level expected by the analytic workflow. These scoped subsets of event data can be distributed into cube *cells*. Then, the event data contained in each cell can be used as input for the analytic workflow (e.g., all the students that took a given course on a given quartile).

The usefulness of the reports is evaluated with dozens of lecturers throughout two evaluation rounds in different academic periods. During the first evaluation round, an initial set of reports was sent to lecturers and feedback was collected through an evaluation form. The feedback was used to restructure the report. Then, a set of restructured reports was sent to lecturers and a group of them were interviewed to asses if the insights contained in the report were better perceived. The results show that, indeed, this is the case.

---

[1] http://fluxicon.com/disco.

[2] http://www.celonis.de/en.

[3] ProM tools is free to download from http://www.promtools.org.

The remainder of this paper is organized as follows. Section 2 discusses related work about educational data analysis and business process reporting. Section 3 provides an overview of the case study and discusses the structure of the reports. Sections 4 and 5 summarize the related work and main concepts related to analytic workflows and process cubes and illustrate how they are concretely applied in this case study. Sections 6 and 7 discusses the reports sent and the results of the two evaluation rounds with the lecturers. Finally, Sect. 8 concludes the paper.

2   Related Work

This section discusses the related work done around business process reporting and educational data analysis. Related work about analytic workflows and process cubes is discussed in Sects. 4 and 5 respectively.

2.1   Business Process Reporting

Business Process Intelligence (BPI) is defined by [3] as the application of Business Intelligence (BI) techniques to business processes. However, behavioral properties of processes (e.g., control-flow) cannot be represented using traditional BI tools. Alternatively, Castellanos et al. [4] provides a broader definition: BPI exploits process information by providing the means for analyzing it to give companies a better understanding of how their business processes are actually executed. It incorporates not only metric-based process analysis, but also process discovery, monitoring and conformance checking techniques as possible ways to understand a process.

Business Process Reporting can be defined as the structured and periodical production of reports containing analysis of process data obtained through BPI techniques.

Business process management suites (e.g., SAP, Oracle) usually provide process reporting capabilities. Often, these process reporting capabilites are an adaptation of general-purpose reporting tools (e.g., Crystal Reports, Oracle Discoverer) to process data [3]. These general-purpose reporting tools are unable to analyze the data from a process perspective (e.g., discover a model).

Most process mining tools (e.g., ProM, Disco) are able to analyze from a process perspective, but they lack reporting and capabilities Others, such as Celonis, offer business process reporting capabilities. However, they are limited to only a few process-perspective analysis components, and each report instance has to be created manually. Furthermore, the event data used as input for the report can only be filtered from the original event data; the granularity level cannot be changed. Also, most of these tools do not allow the comparison of process variants (e.g., students with different grades).

Given the limitations described above, in this paper we used a combination of process mining, analytic workflows and process cubes to provide fully automated process-oriented reports.

2.2 Educational Data Analysis

The analysis of educational data and the extraction of insights from it is related to two research communities: *Educational Data Mining* and *Learning Analytics*.

Educational data mining (EDM) is an emerging interdisciplinary research area that deals with the development of methods to explore data originating in an educational context. EDM uses computational approaches to analyze educational data in order to study educational questions [5,6]. For example, knowledge discovered by EDM algorithms can be used not only to help teachers to manage their classes, understand their students' learning processes and reflect on their own teaching methods, but also to support a learner's reflections on the situation and provide feedback to learners. An extensive survey on the state of the art of EDM is presented in [5].

Learning analytics (LA) is defined by [7] as the measurement, collection, analysis and reporting of data about learners and their contexts, for purposes of understanding and optimising learning and the environments in which it occurs. According to [8], the difference between EDM and LA is that they approach different challenges driving analytics research. EDM focuses on the technical challenge (e.g., How can we extract value from these big sets of learning-related data?), while LA focuses on the educational challenge (e.g., How can we optimise opportunities for online learning?). A discussion on the differences, similarities and collaboration opportunities between these two research communities is presented in [9].

Several process mining techniques (e.g., Fuzzy Miner [10]) have been applied successfully in the context of EDM [11] for analyzing study curriculums followed by students. Notably, the work introduced by [12] aims to obtaining better models (i.e., in terms of model *quality*) for higher education processes by performing data preprocessing and semantic log purging steps. However, most of these techniques are not suitable for analyzing video lecture usage by students, given the inherent lack of structure of such processes.

In this paper, we will use existing and new process mining techniques to obtain insights of student behavior from an educational point of view.

3 A Case Study in Education

Eindhoven University of Technology provides video lectures for many courses for study purposes and to support students who are unable to attend face-to-face lectures for various reasons. Student usage of video lectures and their course grades are logged by the University's IT systems. The purpose of this case study is to show how raw data extracted from the University's IT systems can be transformed into reports that show insights about students' video lecture usage and its relation with course grades by using process mining, process cubes and analytic workflows. Figure 1 describes the overview of this case study.

The data used in this case study contains *video lecture views*, *course grades* and *personal information* of students of the University. Each student and course has a unique identifier code (i.e., *student id, course code*). The data reveals

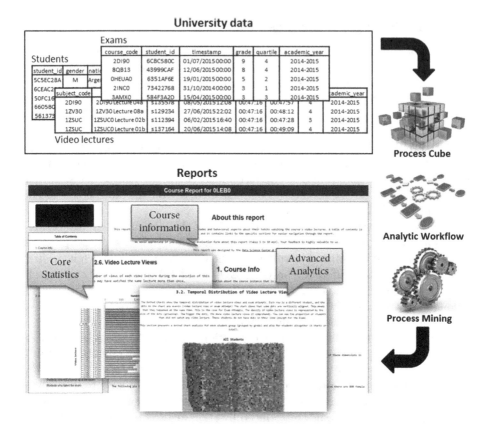

Fig. 1. Overview of the case study: University data is transformed into reports by using process mining, process cubes and analytic workflows.

enormous variability; e.g., thousands of students watch video lectures for thousands of courses and every course has a different set of video lectures, and they have different cultural and study backgrounds, which leads to different behavior. Therefore, we need to provide different reports and, within a report, we need to perform a comparative analysis of the students when varying the grade.

Before describing our approach and the ingredients used, we sketch the report we aim for. The report is composed of three sections: *course information, core statistics* and *advanced analytics*, as shown in Fig. 1.[4] The analysis results refer to all students who registered for the course exam, independently whether or not they participated in it.

---

[4] An example report, where student information has been anonymized, can be downloaded from https://www.dropbox.com/s/565zz94rdo6gg2r/report.zip?dl=0.

The course information section provides general information, such as the course name, the academic year, the number of students, etc. The core statistics section provides aggregate information about the students, such as their gender, nationality, enrolled bachelor or master program, along with course grades distribution and video lecture views. The advanced analytics section contains process-oriented diagnostics obtained through process mining techniques.

The next two sections show how the desired reports can be generated.

4   Analytic Workflows as a Means to Automate Analysis

Process mining experiments usually require analysts to perform many analysis steps in a specific order. As mentioned in Sect. 1, it is not unusual that the same experiment has to be carried out multiple times. This is normally handled by manually executing the analysis steps of the experiment, thus requiring large periods of time and resources and introducing the risk of human-induced errors.

Current process mining tools are not designed to automatically repeat the application of the same process-mining analyses on an arbitrary number of (sub sets of) event logs. Therefore, it is not possible to automatically generate any arbitrary number of reports.

Analytic workflows can be used to address this problem. They are defined by chaining analysis and data-processing steps, each of which consumes input produced by previous steps and generates output for the next steps. Analytic workflows are a specialization of *scientific workflows* tailored towards analytic purposes. Scientific workflows have successfully been applied in many settings [13,14]. The work presented in [15] illustrate the formalization and operationalization of a framework to support process-mining analytic workflows where the steps are linked to the application of process-mining techniques.

In this paper, we combine process mining with analytic workflow systems, which allow one to design, compose, execute, archive and share workflows that represent some type of analysis or experiment. Each activity/step of an analytic workflow is one of the steps to conduct a non-trivial process-mining analysis, which can range from data filtering and transformation to process discovery or conformance checking. Once an analytic workflow is configured, it can be executed with different process data as many times as needed without reconfiguration.

In our case study, for automatically generating the course reports we used RapidProM [15,16], which extends the RapidMiner analytic workflow tool with process mining techniques.[5]

Figure 2a illustrates the analytic workflow that is used to generate each report. Figure 2b shows the explosion of the "Sequence Models" section of the analytic workflow.

The operators shown in Fig. 2 are used for different purposes: *Multipliers* allow one to use the output of an operator as input for many operators. *Filter* operators select a subset of events based on defined criteria. *Process mining*

---

[5] Free version and installation instructions for RapidMiner and the RapidProM extension are available at http://www.rapidprom.org or at the RapidMiner *Marketplace*.

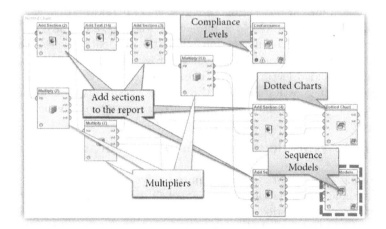

(a) Advanced Analytics section sub-workflow

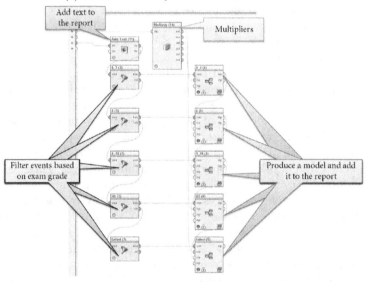

(b) Explosion of the "Sequence Models" sub-workflow

Fig. 2. Implemented analytic workflow used to generate the reports. Each instance of a course can be automatically analyzed in this way resulting in the report described. (Color figure online)

operators are used to produce analysis results. For example, the operators highlighted in blue in Fig. 2b produce a sequence model from each filtered event data.

The complete RapidProM implementation of the analytic workflow used in this case study is available at https://www.dropbox.com/s/g9spsziyv55vsro/single.zip?dl=0. Readers can execute this workflow in RapidMiner to generate a report using the sample event log available at https://www.dropbox.com/s/r3gczshxqxh6a6d/Sample.xes?dl=0.[6]

5 Process Cubes as a Means to Select and Scope Event Data

Processes are not static within moderns organizations but their instances continuously adapt to the dynamic context requirements of modern organizations. Therefore, an event log records executions of several process variants, whose behavior depends on context information (e.g., different courses that may contain different behavior). As a consequence, the event log needs to be split into sub-logs (i.e., one for each variant), each containing all the events that belong to that variant. The naive approach consists on manually filtering the event data. Naturally, this approach is unpractical in scenarios where many different process variants exist.

Process cubes [17] are used to overcome this issue: in a process cube, events are organized into cells using different dimensions. The idea is related to the well-known notion of OLAP (Online Analytical Processing) data cubes and the associated operations, such as slice, dice, roll-up, and drill-down. By applying the correct operations, each cell of the cube contains a sub-set of the event log that complies with the homogeneity assumption mentioned above. This allows one to isolate and analyze the different variants of a process.

Several approaches provide these capabilities, such as [18], which presents an exploratory view on the application of OLAP operations over events. Other process-cube approaches have been applied in specific contexts [19,20]. The term *process cube* was introduced and formalized in [17] with a working prototype presented in [21], and later improved and implemented in [22].

5.1 Basic Concepts

A process cube is characterized by a set of *dimensions*, each of which is associated with one or a group of event's data properties. For each combination of values for the different dimensions, a cell exists in the process cube. Hence, each process-cube cell contains the events that assign certain values to the data properties. Each cell of the process cube contains event data that can be used by process mining techniques. Please note that certain dimensions may be considered as irrelevant and, therefore, they are ignored and are not *visible* in the cube. Also, some dimensions may be not readily available in the event data; however, they can be derived from the existing dimensions. For example, the "Year" and "Day" dimensions can be derived from the "Timestamp" dimension.

The *slice* operation selects a subset of values of a dimension while removing that dimension from the analysis. For example, if the "Year" dimension is sliced

---

[6] When running the workflow, make sure that the *Read File* operator points to the sample event log and the "HTML output directory" parameter of the *Generate Report* operator points to the desired output folder.

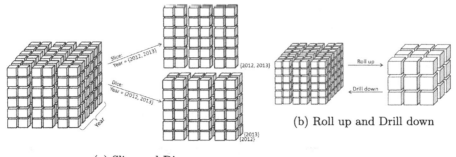

(a) Slice and Dice

(b) Roll up and Drill down

Fig. 3. Schematic examples of cube operations

for Year = {2012, 2013}, only the events in those years are retained. Also, the "Year" dimension is removed from the cube as shown in Fig. 3a. The latter implies that cells with different values for the "Year" dimension and the same values for the other dimensions are merged.

The *dice* operation is similar to the *slice* operation, with the difference that the dicing dimension is retained. So, the dice operation is only removing cells without merging any cells: the dicing dimension can still be used for further exploration of the event data, as shown in Fig. 3a.

The *roll up* and *drill down* operations change the granularity level of a dimension. As shown in Fig. 3b, if a dimension is rolled up, an attribute with a more coarse-grained granularity will be used to create the cells of the cube, and if a dimension is drilled down, an attribute with a more fine-grained granularity will be conversely used. For example, the "Day" dimension can be rolled up to "Month", and the "Month" dimension can be drilled down to "Day".

5.2   Application to the Case Study

For performing process cube operations over the University data we used the *Process Mining Cube* (PMC) tool introduced in [22]. As mentioned before, the starting point is an event data set. This event data set has been obtained by defining and running opportune joins of tables of the database underlying the video-lecture system of the University (see Sect. 3). A fragment of the event data set is shown in Table 1.

Using the event data, we created a process cube with the following dimensions: Student Id, Student Gender, Student Nationality, Student Education Code, Student Education Phase, Course Code, Course Department, Activity, Activity Type, Grade, Timestamp, Quartile and Academic Year.

After *slicing* and *dicing* the cube, thousands of cells are produced: one for each combination of values of the "Course Code", "Quartile" and "Course Grade" dimensions. Each cell corresponds to an event log that can be analyzed using process mining techniques.

Table 1. A fragment of event data generated from the University's system: each row corresponds to an event.

Ev. Id	Student Id	Nat.	Ed. Code	Course code	Activity	Quartile	Acad. year	Timestamp	Grade	⋯
1	1025	Dutch	BIS	2II05	Lecture 1	1	2014–2015	03/09/2012 12:05	6	⋯
2	1025	Dutch	BIS	2II05	Lecture 2	1	2014–2015	10/09/2012 23:15	6	⋯
3	1025	Dutch	BIS	1CV00	Lecture 10	3	2014–2015	02/03/2012 15:36	7	⋯
4	2220	Spanish	INF	1CV00	Lecture 1	3	2014–2015	20/03/2013 16:24	8	⋯
5	1025	Dutch	BIS	2II05	Exam	2	2014–2015	13/12/2012 12:00	6	⋯
6	2220	Spanish	INF	1CV00	Lecture 4	3	2014–2015	25/03/2013 11:12	8	⋯
7	2220	Spanish	INF	1CV00	Exam	3	2014–2015	04/04/2013 12:00	8	⋯
⋯	⋯	⋯	⋯	⋯	⋯	⋯	⋯	⋯	⋯	⋯

We applied our approach that combines process mining, analytic workflows and process cubes to the case study presented in Sect. 3 in two evaluation rounds. The following sections describe the work, reports, results and the feedback obtained on each round.

6   Initial Report

The first evaluation round was conducted in August 2015 and it used the event data corresponding to the academic year 2014–2015. The data used in this round contains **246.526** *video lecture views* and **110.056** *course grades* of **8.122** students, **8.437** video lectures and **1.750** courses. Concretely, we automatically generated a total of **8.750** course reports for 1750 courses given at the University in each of the 5 quartiles (i.e., 4 normal quartiles + *interim* quartile) of the academic year 2014–2015. For reliability of our analysis, we only selected the reports of courses where, on average, each student watched at least 3 video lectures. In total, 89 courses were selected and their reports were sent to the corresponding lecturers.

Section 6.1 shows the first report structure through an example of the reports sent to lecturers in this evaluation round. It also provides a detailed analysis of the findings that we could extract from the report for a particular course. Along with the report, we also sent an evaluation form to the lecturers. The purpose of the evaluation forms is to verify whether lecturers were able to correctly interpret the analysis contained in the report. The results obtained in the first evaluation round are discussed in Sect. 6.2.

6.1   Structure of the Report

To illustrate the structure, contents and value of the reports, we selected an example course: "Introduction to modeling - from problems to numbers and back" given in the third quartile of the academic year 2014–2015 by the Innovation Sciences department at the University. This course is compulsory for all first-year students from all programs at the University. In total, 1621 students attended this course in the period considered. This course is developed in a "flipped classroom" setting, where students watch online lectures containing the course topics and related contents, and in the classroom, they engage these topics in practical settings with the guidance of the instructor.

The video lectures provided for this course are mapped onto weeks (1 to 7). Within each week, video lectures are numbered to indicate the order in which students should watch them (i.e., 1.1 correspond to the first video lecture of the first week). As indicated by the course's lecturer, the first video lectures of each week contain the course topics for that week, and the last video lectures of each week contain complementary material (e.g., workshops, tutorials). The number of video lectures provided for each week depends on the week's topics and related activities, hence, it varies.

Students's behavior can be analyzed from many perspectives. As mentioned in Sect. 2.2, several process mining techniques have been applied in the context of educational data analysis [11].

Initially, we applied traditional process model discovery techniques (e.g., Fuzzy Miner [10], ILP Miner [23], Inductive Visual Miner [24]) to the educational data. However, given the unstructured nature of this data (i.e., students watching video lectures), the produced models were very complex (i.e., *spaghetti* or *flower* models) and did not provide clear insights. Therefore, we opted for other process mining techniques that could help us understand the behavior of students:

Figure 4(a) shows for each video lecture the number of students that watched it. We can observe that the number of students that watch the video lectures decreases as the course develops: most students watched the video lectures corresponding to the first week (i.e., 1.X) but less than half of them watched the video lectures corresponding to the last week (i.e., 7.X). Note that within each week, students tend to watch the first video lectures (i.e., X.1, X.2) more than the last ones (i.e., X.5, X.6). This was discussed with the course's lecturer. It is explained by the fact that, as mentioned before, the first video lectures of each week contain the topics, and the last ones contain complementary material.

Figure 4(b) shows for each student group (i.e., grouped by their grade) the level of conformance, averaged over all students in that group, of the real order in which students watch video lectures, compared with the "natural" or logical order, namely with watching them in sequence (i.e., from 1.1 to 7.4). The conformance level of each student is measured as the *replay fitness* of the data over a process model that contains only the "natural" sequential order. The replay fitness was calculated using conformance checking techniques [25]. We can observe that students with higher grades have higher levels of conformance than students with lower grades.

Figure 4(c) shows the grade distribution for this course where each bar is composed by two parts corresponding to the number of students who watched at least one (red part) video lecture and the number of students who did not (blue part). We can observe that the best students (i.e., with a grade of 8 or above) use video lectures. On the other hand, we observe that watching video

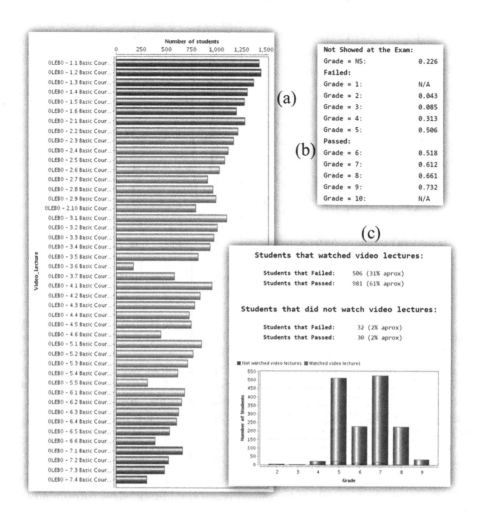

Fig. 4. Analysis results contained in the report of the course 0LEB0: (a) Number of students that watched each video lecture (b) Conformance with the natural viewing order by course grade (c) Grades distribution for students who watched video lectures (in red) or did not (in blue) (Color figure online)

lectures does not guarantee that the student will pass the course, as shown in the columns of students that failed the course (i.e. grade ≤5).

Figure 5 shows dotted charts [26] highlighting the temporal distribution of video-lecture watching for two student groups: (a) students that failed the course with a grade of 5, and (b) students that passed the course with a grade of 6 or 7. Each row corresponds to a student and each dot in a row represents that student watching a video lecture or taking the exam. Note that both charts show a gap where very few video lectures were watched, which is highlighted in the pictures

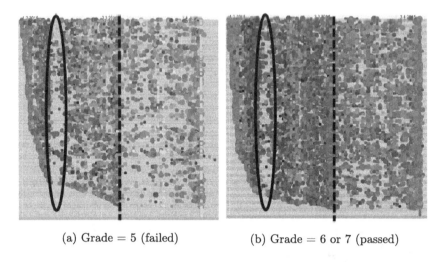

(a) Grade = 5 (failed)   (b) Grade = 6 or 7 (passed)

Fig. 5. Dotted charts for students grouped by their course grades

through an oval. This gap coincides with the *Carnaval* holidays. We can observe that, in general, students that failed watched fewer video lectures. Also note that in Fig. 5(a) the density of events heavily decreases after the mid-term exam (highlighted through a vertical dashed line). This could be explained by students being discouraged after a bad mid-term result. This phenomenon is also present in (b), but not equally evident. We can also observe that most students tend to constantly use video lectures. This is confirmed by the low number of students with only a few associated events.

Figure 6 shows sequence analysis models that, given any ordered sequence of activities, reflects the frequency of directly-follows relations[7] as percentage annotations and as the thickness of edges. The highest deviations from the ordered sequence order are highlighted in colored edges (i.e., black edges correspond to the natural order). This technique was tailored for the generation of reports and it is implemented using a customized RapidProM extension. When comparing (a) students that passed the course with a grade of 6 or 7 with (b) students that had a grade of 8 or 9, we can observe that both groups tend to make roughly the same deviations. Most of these deviations correspond to specific video lectures being skipped. These skipped video lectures correspond in most cases to complementary material. In general, one can observe that the thickness (i.e., frequency) of the arcs denoting the "natural" order (i.e., black arcs) is higher for (b), i.e., those with higher grades. Note that at the beginning of each week we can observe a *recovery* effect (i.e., the frequencies of the natural order tend to increase).

---

[7] The frequency of directly-follows relations is defined for any pair of activities (A, B) as the ratio between the number of times that B is directly executed after A and the total number of times that A is executed.

Table 2. Summary of the classification of statement evaluations performed by lecturers

Statement evaluation	Core statistics section	Advanced analytics section			Sub Total	Total (%)
		Conformance	Temp. Dist	Seq. Analysis		
Correct	261	30	67	32	**390 (89%)**	**61%**
Incorrect	28	5	8	6	**47 (11%)**	
Unknown	95	61	69	58	**283**	**39%**

6.2 Lecturers Evaluation

In addition to the qualitative analysis for some courses like such as the course analyzed in Sect. 6.1, we have also asked lecturers for feedback through an evaluation form linked to each report.[8] The evaluation form provided 30 statements about the analysis contained in the reports (e.g., "Higher grades are associated with a higher proportion of students watching video lectures", "Video lecture views are evenly distributed throughout the course period"). Lecturers evaluated each statement on the basis of the conclusions that they could draw from the report. For each of the 30 statements, lecturers could decide if they agreed or disagreed with the statement, or, alternatively, indicate that they could not evaluate the statement (i.e., "I don't know").

In total, 24 of the 89 lecturers answered the evaluation form. Out of the 720 (24×30) possible statement evaluations, 437 statements were answered with "agree" or "disagree". The remaining cases in which the statement could not be evaluated can be explained by three possible causes: the statement is unclear, the analysis is not understandable, or the data shows no conclusive evidence.

In the case that a statement was evaluated with "agree" or "disagree", we compared the provided evaluation with our own interpretation of the same statement for that report and classified the response as *correct* or *incorrect*. In the case that a statement was not evaluated, the response was classified as *unknown*.

Table 2 shows a summary of the response classification for each section of the report. In total, **89%** of the statement evaluations were classified as *correct*. This indicates that lecturers were capable to correctly interpret the analysis provided in the reports. Note that the *Conformance* section had the highest rate of *unknown* classifications (63.5%). This could be related to understandability issues of the analysis presented in that section.

The evaluation form also contained a few general questions. One of such questions was: "Do you think this report satisfies its purpose, which is to provide insights about student behavior?", for which 7 lecturers answered "yes", 4 lecturers answered "no" and 13 lecturers answered "partially". All the lecturers that responded "partially" provided written feedback indicating the improvements

---

[8] The evaluation form is available at https://www.dropbox.com/s/09y4ypklt70y6d9/form.zip?dl=0.

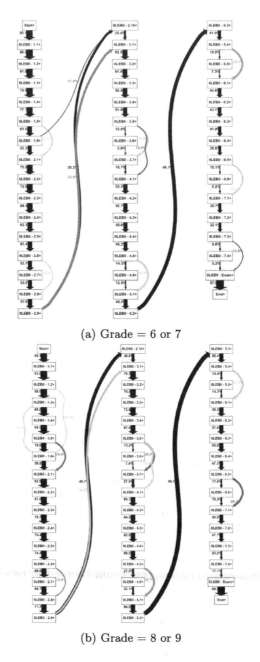

(a) Grade = 6 or 7

(b) Grade = 8 or 9

Fig. 6. Sequence analysis for students grouped by their course grades (Color figure online)

they would like to see in the report. Some of the related comments received were: "It would be very interesting to know if students: (a) did NOT attend the lectures and did NOT watch the video lectures, (b) did NOT attend the lectures, but DID watch the video lectures instead, (c) did attend the lectures AND watch the video lectures too. This related to their grades", "The report itself gives too few insights/hides insights", "It is nice to see how many students use the video lectures. That information is fine for me and all I need to know", and "I would appreciate a written explanation together with your diagrams, next time". Another question in the evaluation form was: "Do you plan to introduce changes in the course's video lectures based on the insights provided by this report?", for which 4 lecturers answered "yes" and 20 answered "no". The results show that the analysis is generally perceived as useful, but that more actionable information is needed, such as face-to-face lecture attendance. However, this information is currently not being recorded by the TU/e. The feedback provided by lecturers was used to improve the report. These improvements are discussed in the next Section.

7   Final Report

We modified the reports based on the feedback obtained in the first evaluation round. The detail of the changes is presented in Sect. 7.1. To assess the quality of the improved report, we conducted a second evaluation round, which was conducted in March 2016 and it used the event data corresponding to the first two quartiles of the academic year 2015–2016. The data used in this round contains **89.936** *video lecture views* and **49.078** *course grades* of **10.152** students, **2.718** video lectures and **1.104** courses. Concretely, we automatically generated a total of **2.208** course reports for 1104 courses given at the University in each of the 2 first quartiles of the academic year 2015–2016. For reliability of our analysis, we only selected the reports of courses where, on average, each student watched at least 3 video lectures. In total, 56 courses were selected and their reports were sent to the corresponding lecturers.

Section 7.1 shows the changes introduced in the report based on the feedback obtained from lecturers in the first evaluation round. It also provides examples of the findings that several lecturers could extract from the report. Along with the report, we also sent an evaluation form to the lecturers. The purpose of the evaluation forms is to verify whether lecturers were able to correctly interpret the analysis contained in the improved report. Unfortunately, in this evaluation round no lecturer answered the evaluation form. Therefore, we held face-to-face meetings with four lecturers, where the results included in the report were discussed. The insights obtained in these meetings are discussed in Sect. 7.2.

7.1   Changes in the Report

According to the feedback obtained from lecturers (reported in Table 2 in Sect. 6.2), the most problematic sections (i.e., highest rate of *unknown* classifications) were the Conformance (63.5% *unknown*) and Sequential Analysis (60.4% *unknown*) sections of the report (described in Sect. 6.1).

Given this feedback and the fact that the interpretation of a specific *replay fitness* value can be misleading for non-process-mining-experts, the conformance section was replaced for a section that describes the *compliance* of students with the "natural" order of watching video lectures based on simpler calculations, defined as follows.

Definition 1 (Compliance Score). *For any given student, their compliance score (CS) w.r.t. the natural order is calculated as* $CS = \sum_{i=1}^{n-1} \frac{df(a_i, a_{i+1})}{count(a_i)}$, *where* $df(a_i, a_{i+1})$ *is the number of times that the student watched lecture* a_{i+1} *directly after* a_i, $count(a_i)$ *is the number of times that the student watched the lecture* a_i *and* n *is the number of video lectures available for the course.*

This new compliance score is easier to interpret: a value of X means that X percent of the video lectures watched by the student were watched in the natural order.

Figure 7 shows an example of the new compliance section of the report. It refers to the course "5ECC0 - Electronic circuits 2" (more details will be given later). Figure 7(a) shows the average compliance scores according to the student's grades, while Fig. 7(b) shows the distribution of students according to their compliance scores.

Regarding the Sequence Analysis section, we simplified the explanatory text of this section in the report. However, this section presents inherent difficulties associated to the analysis of process models: most lecturers are not familiar with process models. Previously, sequence models only referred to frequency deviations. In this round, we decided to incorporate sequence models that show performance information. In these sequence models, an arc indicates the time between the start of the source activity (i.e., video lecture or exam) and the start of the target activity. From these models, one can observe if a given lecture is being fully watched, or if students are skipping most of it after watching a few minutes. Figure 8 shows an example of a sequence model annotated with performance information. This model was obtained from one of the course reports (7U855 - Research methods for the built environment) sent in this evaluation round. From these models we can get interesting insights about the students' behavior on this course. For example, in Fig. 8(a) (i.e., students that obtained a 6 or a 7 in the exam) the arrow between *Lecture 01* and *Lecture 02* states that students that watched *Lecture 02* directly after *Lecture 01*, started watching *Lecture 02* 14 s (in average) after started watching *Lecture 01*. However, in Fig. 8(b) (i.e., students that obtained a 8, 9 or 10 in the exam) this specific behavior is not observed.

7.2   Lecturers Evaluation

As mentioned before, from the 56 reports sent to lecturers in this evaluation round, we obtained no responses to the corresponding evaluation forms. Therefore, we held face-to-face meetings with four lecturers from different departments

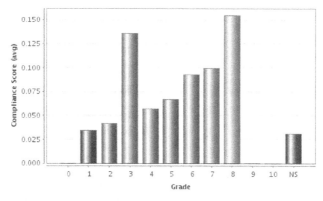

(a) Average student compliance with the "natural" order according to the student's grades

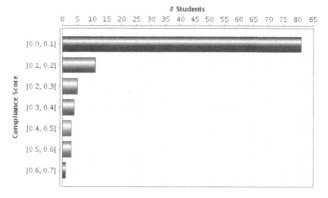

(b) Student distribution over compliance level by range.

Fig. 7. New compliance section of the report for an example course (5ECC0 - Electronic circuits 2)

of the University to discuss the report in general, and to evaluate if the changes introduced in this evaluation round did actually improve the understandability of the report.

In the remainder of this section, we summarize the insights obtained by lecturers when discussing the reports in the face-to-face meetings.

The first lecturer we met was responsible for the course 1CV00 - Deterministic Operations Management, provided by the Industrial Engineering department. In this course, lectures are grouped by topic (i.e., 2 lectures per topic) and topics are independent from each other. Figure 9(a) shows the distribution of students according to their compliance scores for this course. In this chart, we can observe that students have a very low compliance score in general, and it had no correlation with grades. The lecturer defined this behavior as "expected" since the course topics are independent. Figure 9(b) shows the dotted chart containing all

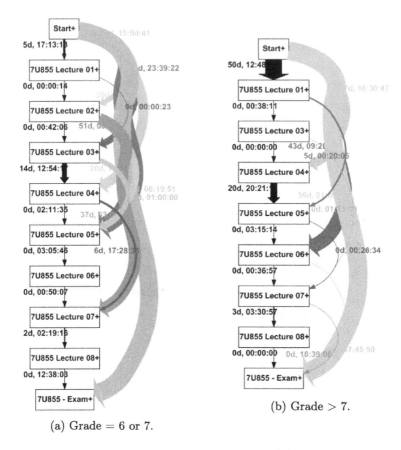

(a) Grade = 6 or 7.

(b) Grade > 7.

Fig. 8. Sequence models annotated with performance information for students grouped by their grade. The models were obtained from the report of course 7U855 - Research methods for the built environment.

the students of the course. Here we can observe two peaks of video lecture usage in weeks 4 and 7 (highlighted with vertical yellow lines), but without context information, we cannot explain why they happened. The lecturer immediately identified these two peaks as the two *mid-term* exams that are part of the course. The interpretation given by the lecturer was that students were using the video lectures to study for these exams. This behavior was expected by the lecturer, but in the past he did not have the information to either confirm or deny it.

The second lecturer was responsible for the course 4EB00 - Thermodynamics, provided by the Mechanical Engineering department. In this course, some topics build on top of knowledge acquired in previous topics, but others are independent. Figure 10(a) shows, for each lecture, the total number of views. We can observe that *Lecture 02a* and *Lecture 05a* had the highest number of views. The lecturer determined that this behavior was expected, since *Lecture 02a* contained most of the definitions and knowledge that students needed to "remember" from

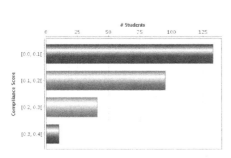

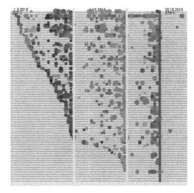

(a) Student distribution over compliance level by range. Compliance scores are relatively low.

(b) Dotted chart for all the students enrolled in the course. Video lecture usage peaks are highlighted with yellow vertical lines.

Fig. 9. Analysis results included in the report of the course 1CV00. (Color figure online)

previous courses. On the other hand, *Lecture 05a* was related to *Entropy*, which was the most difficult topic of the course for students. Figure 10(b) shows the average student compliance with the "natural" order according to the student's grades. We can observe that there is a negative correlation between the compliance scores and the grades. According to the lecturer: "A possible explanation of this could be that students with bad grades could have skipped face-to-face lectures and then needed to watch all the video lectures, while good students attended face-to-face lectures and only watched some video lectures if they needed to clarify something".

The third lecturer was responsible for the course 5ECC0 - Electronic Circuits 2, provided by the Electrical Engineering department. In this course, all the topics were related, every topic built-up on the previous one. Figure 7(a) showed the average student compliance with the "natural" order according to the student's grades. We can observe a positive correlation between compliance scores and grades. The lecturer was positively surprised by this finding, but he considered that the correlation was not strong. Figure 11 shows a fragment of the sequence model with frequency deviations for two different groups of students of the course (i.e., those with a grade lower than 5, and those with a grade equal to 6 or 7). We can observe in Fig. 11(a) that *Lecture 01c* is being skipped by 13% of the students that watched the *Lecture 01b*. This behavior does not occur for students with higher grades (shown in Fig. 11(b)). The lecturer then considered this finding as "unexpected, but positive", since *Lecture 01c* consists of the basic topics from the previous course (i.e., Electronic Circuits 1) and it was meant to refresh student's knowledge. According to the lecturer, the fact that students did not need to watch it is positive.

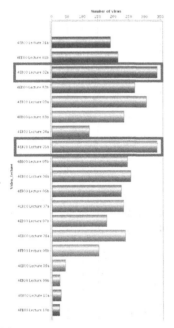

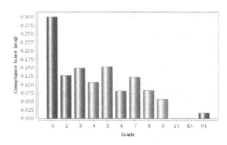

(b) Average student compliance with the "natural" order according to the student's grades. There seems to be a negative correlation between compliance scores and grades.

(a) Number of views of each video lecture of the course. Lectures 02a and 05a were the most watched by students (highlighted in red).

Fig. 10. Analysis results included in the report of the course 4EB00. (Color figure online)

The fourth lecturer was responsible for the course 5XCA0 - Fundamentals of Electronics, provided by the Electrical Engineering department. This course considers topics are relatively independent from each other. Figure 12(a) shows, for each lecture, the total number of views. It is interesting to notice that the *Lecture 05a*, the video lecture most watched by students (highlighted in red) is an *instruction* lecture (i.e., a lecture that consists of exercises instead of topics). Given this finding, the lecturer has expressed the intention of splitting that video lecture, for the next executions of the course, into a series of 10-minute web lectures comprehending all the different types of exercises covered in the video lecture. Figure 12(b) shows the student distribution over ranges of compliance score. It is clear from the chart that most students have a very low compliance score w.r.t. the "natural" viewing order. This was justified by the lecturer through the following statement: "The topics are relatively disconnected, and it seems that most students would watch only specific lectures".

The general comments that we received from lecturers are summarized as follows. "It would be interesting to see the correlation with face-to-face lectures to see if students use video lectures as a replacement or as a complement for

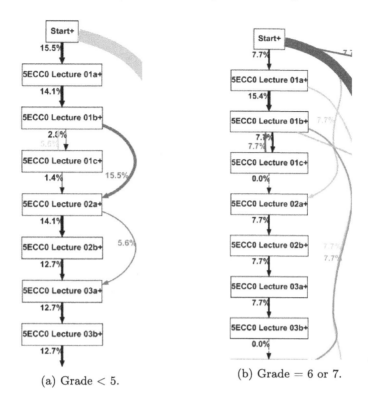

(a) Grade < 5.
(b) Grade = 6 or 7.

Fig. 11. Fragment of the sequence model with frequency deviations for all students. In (a), Lecture 1c is being skipped. These charts were included in the report of the course 5ECC0 - Electronic Circuits 2.

them". "Video lectures are very good for the *middle* students. good students do not seem to need them as much". "I should split the most visited video lectures into a series of web lectures (i.e., 10 min recordings of specific topics) so I could really know which topics are the most difficult for the students". "Students tend to use exercise lectures much more intensively than the actual theory. They seem to be exam-oriented, as they prepare mostly watching exercises".

Regarding the report itself, again we received suggestions to incorporate face-to-face lecture attendance. As mentioned in Sect. 6, it is very difficult to record face-to-face attendance of students for technical reasons. Other lecturers suggested that we incorporate student feedback into the report. We certainly recognize the potential that incorporating the students' feedback on the report could have in the insights that the lecturer can obtain from it. We plan to incorporate this feedback and its potentially positive effects in the reports for the next quartile.

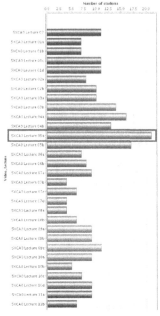

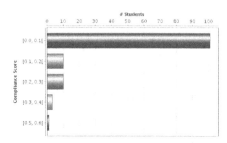

(b) Student distribution over compliance level by range. Most students have a very low compliance score (i.e., between 0% and 10%).

(a) Number of views of each video lecture of the course. Lecture 05a was the most watched by students (highlighted in red).

Fig. 12. Analysis results included in the report of the course 5XCA0. (Color figure online)

8   Conclusion

This paper has illustrated the benefits of combining the complementary approaches of process cubes and analytic workflows in the field of process mining. In particular, the combination is beneficial when process mining techniques need to be applied on large, heterogenous event data of multidimensional nature.

To demonstrate such benefits, we applied the combined approach in a large scale case study where we provide reports for lecturers. These reports correlate the grades of students with their behavior while watching the available video lectures. We evaluated the usefulness of the reports in two evaluation rounds. The second evaluation round presented an improved report, which was modified based on the feedback obtained in the first evaluation round. Unlike existing *Learning Analytics* approaches, we focus on dynamic student behavior. Also, descriptive analytics would not achieve similar analysis results because they do not consider the process perspective, such as the ordering of watching video lectures.

Educational data has been analyzed by some disciplines in order to understand and improve the learning processes [5–9], even employing process cubes [20]. However, these analyses were mostly focused on individual courses. No research work has previously been conducted to allow large-scale process mining analysis where reports are automatically generated for any number of courses. Our approach has made it possible by integrating process mining with analytic workflows, which have been devised for large-scale analysis, and process cubes, which provide the capabilities needed to perform comparative analyses.

As future work, the report generation will be extended to *Massive Open Online Courses* (MOOCs) given by Eindhoven University of Technology. This type of courses are particularly interesting due to the fact that face-to-face lectures are not used: video lectures are the main channel used by students for accessing the course topics. For example, over **100.000** people from all over the world registered for the two executions of the MOOC *Process Mining: Data science in Action*.[9] We also plan to apply this analysis to the courses provided by the *European Data Science Academy* (EDSA).[10]

References

1. van der Aalst, W.M.P.: Process Mining: Discovery, Conformance and Enhancement of Business Processes, 1st edn. Springer, Heidelberg (2011)
2. Dongen, B.F., Medeiros, A.K.A., Verbeek, H.M.W., Weijters, A.J.M.M., van der Aalst, W.M.P.: The ProM framework: a new era in process mining tool support. In: Ciardo, G., Darondèau, P. (eds.) ICATPN 2005. LNCS, vol. 3536, pp. 444–454. Springer, Heidelberg (2005). doi:10.1007/11494744_25
3. Grigori, D., Casati, F., Castellanos, M., Dayal, U., Sayal, M., Shan, M.C.: Business process intelligence. Comput. Ind. **53**(3), 321–343 (2004). Process/Workflow Mining
4. Castellanos, M., Alves de Medeiros, A.K., Mendling, J., Weber, B., Weijers, A.J.M.M.: Business process intelligence. In: Cardoso, J., van der Aalst, W.M.P. (eds.) Handbook of Research on Business Process Modeling, pp. 456–480. IGI Global, Hershey, PA, USA (2009)
5. Romero, C., Ventura, S.: Educational data mining: a review of the state of the art. IEEE Trans. Syst. Man Cybern. Part C Appl. Rev. **40**(6), 601–618 (2010)
6. Gorissen, P.J.B.: Facilitating the use of recorded lectures: analysing students' interactions to understand their navigational needs. Ph.D. thesis, Eindhoven University of Technology (2013)
7. Siemens, G.: Learning analytics: the emergence of a discipline. Am. Behav. Sci. **57**(10), 1380–1400 (2013)
8. Ferguson, R.: Learning analytics: drivers, developments and challenges. Int. J. Technol. Enhanced Learn. **4**(5–6), 304–317 (2012)
9. Siemens, G., Baker, R.S.J.d.: Learning analytics and educational data mining: towards communication and collaboration. In: Proceedings of the 2nd International Conference on Learning Analytics and Knowledge, LAK 2012, pp. 252–254. ACM, New York (2012)

---

[9] http://www.coursera.org/course/procmin.

[10] http://edsa-project.eu.

10. Günther, C.W., van der Aalst, W.M.P.: Fuzzy mining – adaptive process simplification based on multi-perspective metrics. In: Alonso, G., Dadam, P., Rosemann, M. (eds.) BPM 2007. LNCS, vol. 4714, pp. 328–343. Springer, Heidelberg (2007). doi:10.1007/978-3-540-75183-0_24

11. Trcka, N., Pechenizkiy, M., van der Aalst, W.M.P.: Process mining from educational data. In: Handbook of educational data mining, pp. 123–142. CRC Press, London (2010)

12. Ly, L.T., Indiono, C., Mangler, J., Rinderle-Ma, S.: Data transformation and semantic log purging for process mining. In: Ralyté, J., Franch, X., Brinkkemper, S., Wrycza, S. (eds.) CAiSE 2012. LNCS, vol. 7328, pp. 238–253. Springer, Heidelberg (2012). doi:10.1007/978-3-642-31095-9_16

13. Jaeger, E., Altintas, I., Zhang, J., Ludäscher, B., Pennington, D., Michener, W.: A scientific workflow approach to distributed geospatial data processing using web services. In: Proceedings of the 17th International Conference on Scientific and Statistical Database Management (SSDBM 2005), Berkeley, CA, USA, pp. 87–90. Lawrence Berkeley Laboratory (2005)

14. Turner, K., Lambert, P.: Workflows for quantitative data analysis in the social sciences. Int. J. Softw. Tools Technol. Transf. **17**(3), 321–338 (2015)

15. Bolt, A., de Leoni, M., van der Aalst, W.M.P.: Scientific workflows for process mining: building blocks, scenarios, and implementation. Int. J. Softw. Tools Technol. Transf. (2015). doi:10.1007/s10009-015-0399-5

16. Mans, R.S., van der Aalst, W.M.P., Verbeek, H.M.W.: Supporting process mining workflows with RapidProM. In: Proceedings of the BPM Demo Sessions 2014 Colocated with the 12th International Conference on Business Process Management (BPM). CEUR Workshop Proceedings, vol. 1295, pp. 56–60. CEUR-WS.org (2014)

17. van der Aalst, W.M.P.: Process cubes: slicing, dicing, rolling up and drilling down event data for process mining. In: Song, M., Wynn, M.T., Liu, J. (eds.) AP-BPM 2013. LNBIP, vol. 159, pp. 1–22. Springer, Heidelberg (2013). doi:10.1007/978-3-319-02922-1_1

18. Ribeiro, J.T.S., Weijters, A.J.M.M.: Event cube: another perspective on business processes. In: Meersman, R., et al. (eds.) OTM 2011. LNCS, vol. 7044, pp. 274–283. Springer, Heidelberg (2011). doi:10.1007/978-3-642-25109-2_18

19. Vogelgesang, T., Appelrath, H.J.: Multidimensional process mining: a flexible analysis approach for health services research. In: Proceedings of the Joint EDBT/ICDT 2013 Workshops, EDBT 2013, pp. 17–22. ACM, New York (2013)

20. van der Aalst, W.M.P., Guo, S., Gorissen, P.: Comparative process mining in education: an approach based on process cubes. In: Ceravolo, P., Accorsi, R., Cudre-Mauroux, P. (eds.) SIMPDA 2013. LNBIP, vol. 203, pp. 110–134. Springer, Heidelberg (2015). doi:10.1007/978-3-662-46436-6_6

21. Mamaliga, T.: Realizing a process cube allowing for the comparison of event data. Master's thesis, Eindhoven University of Technology, Eindhoven, The Netherlands (2013)

22. Bolt, A., van der Aalst, W.M.P.: Multidimensional process mining using process cubes. In: Gaaloul, K., Schmidt, R., Nurcan, S., Guerreiro, S., Ma, Q. (eds.) CAISE 2015. LNBIP, vol. 214, pp. 102–116. Springer, Heidelberg (2015). doi:10.1007/978-3-319-19237-6_7

23. Werf, J.M.E.M., Dongen, B.F., Hurkens, C.A.J., Serebrenik, A.: Process discovery using integer linear programming. In: Hee, K.M., Valk, R. (eds.) PETRI NETS 2008. LNCS, vol. 5062, pp. 368–387. Springer, Heidelberg (2008). doi:10.1007/978-3-540-68746-7_24

24. Leemans, S.J.J., Fahland, D., van der Aalst, W.M.P.: Exploring processes and deviations. In: Fournier, F., Mendling, J. (eds.) BPM 2014. LNBIP, vol. 202, pp. 304–316. Springer, Heidelberg (2015). doi:10.1007/978-3-319-15895-2_26

25. van der Aalst, W.M.P., Adriansyah, A., van Dongen, B.F.: Replaying history on process models for conformance checking and performance analysis. Wiley Interdisc. Rev. Data Min. Knowl. Discovery **2**(2), 182–192 (2012)

26. Song, M., van der Aalst, W.M.P.: Supporting process mining by showing events at a glance. In: Proceedings of the 17th Annual Workshop on Information Technologies and Systems (WITS), pp. 139–145 (2007)

Detecting Changes in Process Behavior
Using Comparative Case Clustering

B.F.A. Hompes[1,2(✉)], J.C.A.M. Buijs[1], Wil M.P. van der Aalst[1], P.M. Dixit[1,2],
and J. Buurman[2]

[1] Department of Mathematics and Computer Science,
Eindhoven University of Technology, Eindhoven, The Netherlands
{b.f.a.hompes,j.c.a.m.buijs,w.m.p.v.d.aalst}@tue.nl
[2] Philips Research, Eindhoven, The Netherlands
{prabhakar.dixit,hans.buurman}@philips.com

Abstract. Real-life *business processes* are complex and often exhibit a high degree of variability. Additionally, due to changing conditions and circumstances, these processes continuously evolve over time. For example, in the healthcare domain, advances in medicine trigger changes in diagnoses and treatment processes. Case data (e.g. treating physician, patient age) also influence how processes are executed. Existing *process mining* techniques assume processes to be static and therefore are less suited for the analysis of contemporary, flexible business processes. This paper presents a novel *comparative case clustering* approach that is able to expose changes in behavior. Valuable insights can be gained and process improvements can be made by finding those points in time where behavior changed and the reasons why. Evaluation using both synthetic and real-life event data shows our technique can provide these insights.

Keywords: Process mining · Trace clustering · Concept drift · Process comparison

1 Introduction

The execution of business processes is typically influenced by many external factors. Due to changing conditions and circumstances, these processes continuously evolve over time. For example, advances in medicine can change how patients are treated or how diagnoses are made in hospitals. Other changing circumstances could be legislation, seasonal effects or even involved resources. As a result, in these flexible processes, many cases follow a unique path through the process. This variability causes problems for existing process mining techniques that assume processes to be structured and in a steady state. Contemporary process mining techniques return spaghetti-like processes and potentially misleading results when this is not the case [3,9,18]. The discovered process models capture behavior possible at any given point in time. Often, however,

P. Ceravolo and S. Rinderle-Ma (Eds.): SIMPDA 2015, LNBIP 244, pp. 54–75, 2017.
DOI: 10.1007/978-3-319-53435-0_3

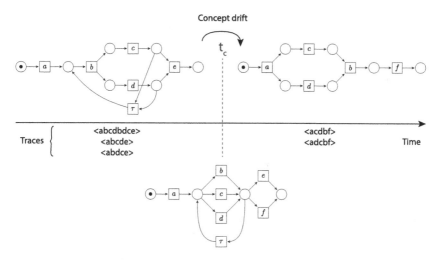

Fig. 1. Demonstration of (sudden) concept drift. The upper process models (Petri nets) accurately describe behavior before and after a change point t_c, whereas the lower model tries to capture all behavior, thereby returning a misleading process model that allows for too much behavior that is not seen in the event log.

the process context changes and what was possible before is no longer possible or vice-versa. Figure 1 shows the relevance of the problem with an example process that has changed over time. This so-called *concept drift* is one of the key challenges in process mining, as discussed in the Process Mining Manifesto [2]. Few techniques have been proposed to deal with concept drift in a business process setting [5,7,11,13,19]. Where existing techniques focus mainly on change in control-flow, our focus is on detecting changing behavior including data aspects.

We identify two types of change for which existing techniques do not work well: individual cases for which behavior changed, and changes in the overall process. The former are typically seen as outliers, because those cases are usually infrequent or dissimilar to the majority of other cases. However, specific changes in context might require cases to change behavior. Alternatively, processes themselves can be subject to change, due to changing conditions and circumstances.

Finding changes in behavior can be of great value to process owners. Detecting unwanted changes, for example, can identify potential risks while positive change can lead to perdurable process improvement. Techniques that consider the entire event log at once cannot show the individual changes that occur throughout the lifetime of a process [5]. For example, specific types of behavior might occur for a limited time only, or can have a seasonal nature. Behavior can also merge with, or emerge from other behavior.

Recently a novel technique for the detection of common and deviating behavior using *trace clustering* was proposed in [10]. Here, the process context is considered by taking both control-flow as well as case and event data into account.

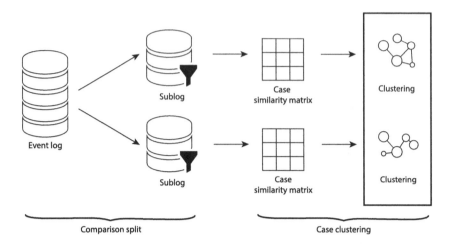

Fig. 2. A graphical overview of the approach. The event log is split into sublogs. Differences in behavior can be compared using comparative case clustering.

Hence, clustering of cases is not limited to finding similar execution paths. In this paper, we extend the work in [10] by providing a novel technique for *change point detection* and a means to compare case clustering results. Figure 2 shows a high-level overview of our approach. Event logs are split up and the behavior of cases in the resulting sublogs is compared. Our technique aids the analyst by providing indications of the points in time where behavior changed and thus where interesting comparisons can be made. The case clustering technique is based on the Markov cluster (MCL) algorithm [17]. It is able to autonomously discover a (non-predefined) number of clusters of different sizes and densities, leading to the separation of mainstream and deviating behavior.

By utilizing the time dimension we can discover temporal evolutions in process behavior. In order to detect change points, we look at how similarities between cases evolve over time. We consider the effect propagated by new events on a clustering of cases as an indicator of changing behavior. We account for the effect of case maturity in similarity calculation using an aging strategy. For example, it might be the case that certain behavior has not been seen for a while or is seasonal of nature. It would be of interest to analysts to discover the points in time where this behavior has returned. As such, in our technique, similarity between cases is corrected for age. Differences in behavior are then analyzed by comparing clusterings created for two selected partitions of the event logs before and after the detected change point. Partitions include different customer types, cases handled by different resources, etc.

The remainder of this paper is organized as follows. Section 2 briefly introduces necessary preliminary definitions. Section 3 describes the case clustering technique on which our method is built. The detection of change points and related considerations are described in Sect. 4. Section 5 describes the comparative trace clustering approach. An experimental evaluation on a synthetic and two real-life event logs is performed in Sect. 6. Section 7 discusses related work. The paper concludes with a summary and planned future work in Sect. 8.

2   Preliminaries

Typically, the executed *events* of multiple *cases* of a *process* are recorded in an *event log*. An event represents one execution of an activity for a case, and potential contains additional data attributes such as a timestamp or the responsible resource. A trace is a finite sequence of events, and describes one specific instance (i.e. a case) of the process at hand in terms of the executed events. A case can also have additional (case-level) attributes such as a patient birthdate or customer type. Definitions for events and cases are based on those in [1].

Definition 1 (Event, attribute). *Let $\mathcal{E}$ be the event universe, i.e. the set of all possible event identifiers. Events may be characterized by various attributes, e.g. an event may have a timestamp, correspond to an activity, be executed by a particular person, etc. Let N be a set of attribute names. For any event $e \in \mathcal{E}$ and attribute name $n \in N$: $n(e)$ is the value of attribute n for event e. If event e does not have an attribute named n, then $n(e) = \perp$ (null value).*

Typically, the following attributes are present in all events: $activity(e)$ is the *activity* associated to event e, $time(e)$ is the *timestamp* of e, and $resource(e)$ is the *resource* associated to e. Additional event attributes can be the cost associated with the event, the outcome of an activity (e.g. diagnosis result), etc.

Definition 2 (Case, trace, event log). *Let $\mathcal{C}$ be the case universe, i.e. the set of all possible case identifiers. Cases, like events, have attributes. For any case $c \in \mathcal{C}$ and attribute name $n \in N$: $n(c)$ is the value of attribute n for case c ($n(c) = \perp$ if c has no attribute named n). Each case has a mandatory attribute 'trace': $trace(c) \in \mathcal{E}^*$. $\hat{c} = trace(c)$ is a shorthand notation for referring to the trace of a case. A trace is a finite sequence of events $\sigma \in \mathcal{E}^*$, such that each event appears only once, i.e. for $1 \leq i < j \leq |\sigma|$: $\sigma_i \neq \sigma_j$. For any sequence $s = \langle s_1, s_2, \ldots, s_n \rangle$, $set(s) = \{s_1, s_2, \ldots, s_n\}$ converts a sequence into a set, e.g. $set(\langle a, b, c, b, c, d \rangle) = \{a, b, c, d\}$. An event log is a set of cases $L \subseteq \mathcal{C}$ such that each event appears at most once in the entire log, i.e. for any $c, c' \in L$ such that $c \neq c'$: $set(\hat{c}) \cap set(\hat{c}') = \emptyset$.*

In the example cases in Figs. 1 and 4 a simplified form is used where events are represented solely by the activities they represent. In this form, a trace is a sequence of activities and an event log a multiset of traces (since in this simplified form cases can share the same sequence of events, and no additional data attributes are present).

In order to automatically separate the event log into multiple sublogs and to detect changes in behavior, we use a time window. Time windows are defined by their length, either in time units or in a specific number of events. This has as an effect that their length in real-time can vary. Behavior in time periods at which less events are recorded (such as low seasons) can thus be represented by a single window, whereas high-frequency periods are divided in more time windows. Here, we use a tumbling time windowing strategy. More detail about the different possible types of time windows is given in Subsect. 4.1.

Definition 3 (Time window). *Let W be a time window (window). A window W has two properties: W_s, and W_e, which denote the start and end times of W. The length of W is denoted by $|W| = W_e - W_s$. $L\lceil_W = L_W \subseteq L$ denotes the projection of an event log $L \subseteq \mathcal{C}$ to window W. In L_W, case information (attribute values, events) known after time W_e in L is removed, i.e. for all cases, the prefix up to W_e of their trace and case information known at time W_e is kept.*

Cases in an event log can be clustered based on multiple perspectives. Perspectives can be based on case and/or event attributes (such as the age of a patient), simple or more advanced control-flow patterns, or can be derived values such as the time spent in the hospital. By using both control-flow and data perspectives the process context is considered.

Definition 4 (Perspective). *Let P be a perspective. $\lceil_P : \mathcal{C} \to \mathbb{R}^m$ denotes the function mapping a case to a real vector of length m according to perspective P. m is the number of attributes in P. For example, m can be the number of different resources in the log or the amount of distinct diagnoses there are. $c\lceil_P$ denotes the projection of case $c \in \mathcal{C}$ to a perspective P. Furthermore, we let $c\lceil_{\{P_1,P_2,...,P_K\}} = c\lceil_{P_1}\|c\lceil_{P_2}\|...\|c\lceil_{P_K}$, i.e. the resulting profile vectors from projection to multiple perspectives are concatenated.*

The Markov cluster algorithm uses a similarity matrix between cases as its input. This matrix holds pair-wise similarity values between the profile vectors obtained by projecting each case to the selected perspectives.

Definition 5 (Case similarity matrix). *Let $L \subseteq \mathcal{C}$ be an event log.*
$\mathcal{M}(L) = (L \times L) \to [0.0, 1.0]$ denotes the set of all possible case similarity matrices over L. For cases $c, c' \in L$ and a case similarity matrix $M \in \mathcal{M}(L)$, $M(c, c')$ denotes the similarity between c and c'.

Similarity values for cases are multiplied by an age factor that decreases with time in order to correct for the similarity between current and older cases in the change detection process. This way, we account for seasonal temporal changes, or temporally infrequent behavior.

Definition 6 (Age vector). *Let $L \subseteq \mathcal{C}$ be an event log. $\vec{a}(L, W) = L \to [0.0, 1.0]$ denotes an age vector a over L, for time window W. For any case $c \in L$, $\vec{a}(c, W)$ denotes the age factor of c for window W.*

Definition 7 (Case clustering). *Let $L \subseteq \mathcal{C}$ be an event log. A case cluster (cluster) over L is a subset of L. A case clustering $TC \subseteq \mathcal{P}(L)^1$ is a set of case clusters over L. We assume every case to be part of at least one cluster, i.e. $\bigcup TC = L$. Cases can be in multiple clusters, i.e. cluster overlap is allowed.*

---

[1] $\mathcal{P}(L)$ denotes the powerset over event log L, i.e. all possible sublogs of L.

3   Case Clustering

Discovering a process model on an entire real-life event log will often lead to a spaghetti model since it has to represent all past traces [3,9,18]. Similarly, clustering the entire event log will show groups of behavior that were possible at any given point in time. The temporal evolution of the behavioral differences captured by the clustering is not shown.

Our change detection technique is based on the technique proposed in [10] where case clustering and outlier detection are combined in order to find mainstream and deviating behavior. This technique relies on the Markov cluster (MCL) algorithm [17] to find clusters of cases that share behavior on a set of selected perspectives. By incorporating both control-flow and case data, the process context is taken into account. MCL was chosen over alternative clustering techniques because of the following properties. The number of clusters is discovered rather than set beforehand, hence changes in behavior will be reflected in change in the clustering. Because MCL is neither biased towards globular or local clusters and is able to find clusters of different density, exceptional cases will not be clustered together with common behavior, i.e. they can be distinguished based on cluster sizes. As a result, new types of cases will result in new clusters rather than to be added to existing clusters. Since cluster overlap is possible, evolution of cases from one cluster to another over time can be detected as well. For more details on differences with alternative clustering approaches the reader is referred to [10].

MCL uses a stochastic similarity matrix between cases as its input. It alternates an expansion step that raises the matrix to a given power with the inflation step. Inflation raises each element to a given power and normalizes the matrix such that it is stochastic again. As such, there are two parameters to MCL: the expansion and inflation parameter, which both influence clustering granularity. This alternation eventually results in the separation of the matrix into different components, which are interpreted as clusters. In our case, the MCL algorithm takes as input a left stochastic version of the case similarity matrix, i.e. the columns are normalized. In order to create a case similarity matrix, cases in the event log are mapped to a profile vector by projecting them to a selected set of perspectives. A pair-wise similarity score is then calculated between these profile vectors. Applying the MCL algorithm to the resulting case similarity matrix gives us a case clustering over the log. The process of applying MCL to case clustering is visualized in Fig. 3.

More formally, for all cases $c \in L \subseteq C$, we project c to our chosen set of perspectives P, i.e. $c\lceil_P$. Next, for each pair of the resulting profile vectors, we compute the pair-wise vector similarity. In this paper we use the cosine similarity, i.e. $M(c, c') = \frac{c\lceil_P \cdot c'\lceil_P}{||c\lceil_P|| \, ||c'\lceil_P||}$, $\forall c, c' \in L$, where $M \in \mathcal{M}(L)$. However, any vector similarity metric can be used. Cosine similarity was chosen because of its proven effectiveness, efficient calculation and boundedness to $[0, 1]$. Also, it is able to represent non-binary term weights and allows for partial matching. A typical downside of vector similarity measures is that the order of terms is lost.

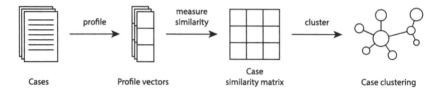

Fig. 3. Overview of the application of the Markov cluster algorithm.

This problem can be solved by incorporating order in the perspectives, such as the occurrence of frequent patterns.

4 Detecting Change in Behavior

In the extreme case, we could cluster the cases in the event log after every new event that arrives, with the aim of finding significant changes in behavior. However, due to the nature of different clustering algorithms (including the Markov cluster algorithm), this is too time and resource consuming. Often, even clustering after a specific number of new events or time units would be too expensive still, or would need excessively large window sizes to be feasible.

The Markov cluster algorithm autonomously discovers a number of clusters with varying sizes and densities based on its input. Over time, the occurrence of new events will change the similarity between cases, which is reflected in the similarity matrix. As this matrix is the input for the clustering algorithm, the impact on the case similarity matrix is a good indicator for how much the clustering will change. Hence, we can use the evolution of this matrix over time as a reliable predictor for change in the clustering output. Therefore, we propose to detect changes in behavior by utilizing the change over time of the case similarity matrix. Similar to the approaches that use statistical tests [5,12,13], differences in the matrix indicate change in behavior. An overview of our approach is illustrated in Fig. 4, where five simple example traces are drawn over time.

4.1 Splitting the Event Log

In order to calculate change in behavior, the events belonging to the different cases are split over several consecutive time windows. Different windowing strategies exist in literature, such as adaptive, tumbling, sliding, and flexible windows. In our case, we use a tumbling windowing technique that facilitates the comparison of cases from their start to the end of the window. For each time window, cases that have events in or before that window are considered in the calculation of the similarity matrix, as is depicted in Fig. 5. Events occurring after the current time window are not considered. As such, for each case, only the known attributes and prefix of its trace are taken into account. Different window sizes can be considered. For example, we can compute a new similarity matrix every

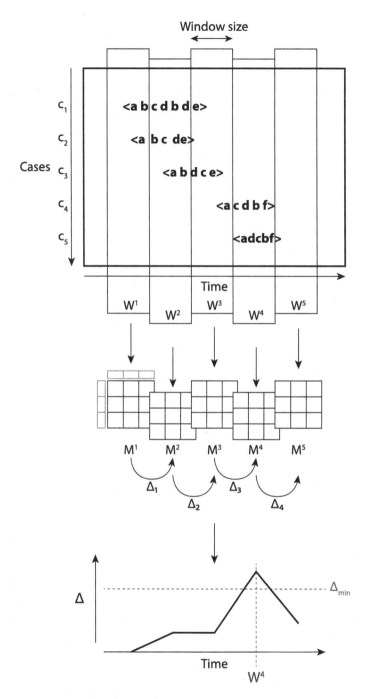

Fig. 4. A graphical overview of the change detection approach. In order to detect changes in behavior, the difference in similarity matrices over time is calculated. In this example, significant change occurred in window W^4.

10 events, or after every 30 min. The choice for sensible window sizes depends entirely on the type of event data that is being analyzed. As such, no rule of thumb can be provided.

Note that it is not possible to use windowing techniques that vary the window size based on the automated detection of changes in the data, since this detection is the goal of our approach. We consider attributes that are not specifically output by events (such as patient age) to be known from the very first event in the trace of a case. As a result, when looking at this type of case-level data, a change point will be shown for the first event of a case. Related research has been performed in the field of streaming data [8], and process mining in a streaming setting in particular [6].

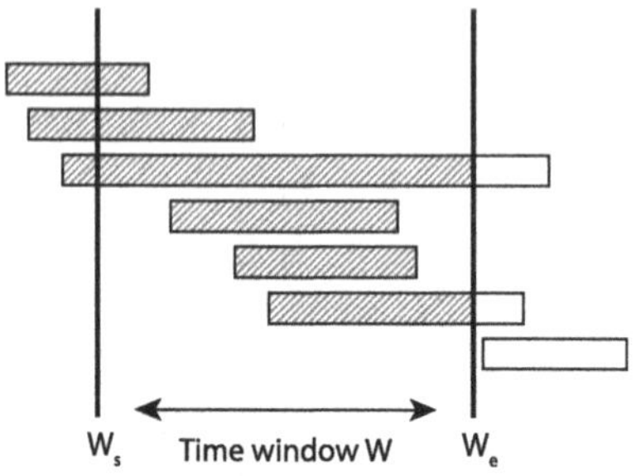

Formally, for every time window W^i, where $i \in [1, n]$, an event log $L \subseteq C$ is projected to W^i, resulting in n sublogs $L\lceil_{W^i}$. Note that $W^i_s = W^{i-1}_e$, for all $1 < i \leq n$, i.e. all windows are consecutive, and there is no overlap in their start and end times. For every W^i, $L\lceil_{W^i}$ contains all events in L up to time W^i_e. Also note that as discussed, $|W^i|$ is not necessarily equal to $|W^{i+1}|$, for example when we choose a number of events as the size of the time windows.

Fig. 5. Selection of cases using a time window. Cases are represented by horizontal bars. All events for all cases up to W_e are taken into account.

4.2 Calculating Case Similarity

After splitting the events in the event log over consecutive time windows, a similarity matrix is calculated for each sublog and compared with the similarity matrix calculated for the sublog that was obtained by projecting the event log to the previous time window. Before this is done, however, we account for case age.

Without correcting for the age of cases, the effect of seasonal temporal behavior patterns are hidden. For example, imagine a process that is executed differently in winter (low season) compared to the other three season. Without accounting for case age, cases that are observed in winter will exhibit a high similarity with those cases observed in winter the previous year. No change will be detected in the transition from fall to winter. Similarly, cases that are exceptional but not unique will become less detectable over time.

In order to correct for the age of cases, we keep a vector of case age factors, that are used to decrease case similarity for those cases further back in history. This factor can be reduced over time in different ways (e.g. exponentially or linearly), in order to consider only recent cases or also the earlier ones. So, for every sublog $L\lceil_{W^i}$, we create a case similarity matrix M^i, as described in Sect. 3. This matrix is corrected for age by multiplying it by $A^i = \sum_{j=1}^{|L\lceil_{W^i}|} \vec{a}(L, W^i)_j \cdot E_j,$

where E_j is the $|L\lceil_{W^i}| \times |L\lceil_{W^i}|$ matrix with a 1 on position (j, j) and zeros everywhere else. In the resulting case similarity matrix only the upper triangle has been accounted for age. We mirror the upper triangle downwards to make the matrix symmetrical again. To exponentially decrease the impact of a case to the case similarity matrix, we multiply the age vector with some aging factor in every new time window, i.e. $\vec{a}(L, W^i) = \vec{a}(L, W^{i-1}) \cdot s$, where $s \in [0, 1]$. Alternatively, in order to decrease the age linearly, we can subtract a value for each element in the vector instead of multiplying with a factor.

Note that this type of correction for age combined with the windowing technique described in Subsect. 4.1 differs in effect from traditional sliding window techniques, since similarity between cases might result from attributes or events that were known before a certain time window. As such, we decrease the impact of the similarity value obtained by comparing all case data known so far, rather than computing similarity only on a subset of that data.

4.3   Computing Change

Once age has been accounted for, the difference in similarity matrices can be calculated. For the calculation of the change in similarity matrices, we look at the change in the matrix values. The maximal difference equals the amount of cells we have in the case similarity matrix of the latest time window divided by two (since the matrix is symmetrical). Hence, we calculate the change value as a percentage of this amount. Formally, the change between two case similarity matrices M^1 and M^2 of size k equals $\frac{2}{k} \times \sum(|M^2 - M^1|)$, where $\sum M = \sum_{c,c'} M(c, c')$. Note that the dimensionality of similarity matrices changes over time due to the cases that are included in the respective time windows. When two matrices of different dimensions are compared, both matrices are extended with empty cells referring to cases that are only present in the other time window.

Interesting change points can be deduced from the evolution of the change value over time. An increase indicates that in this window, more things changed compared to in the previous window. A drop indicates less change. Accordingly, interesting change points are those points in time where the change value exhibits spikes. Once changes in behavior have been discovered, clusterings can be created in order to compare behavior before and after the identified change points.

For example, take the cases depicted in Fig. 4, and consider cases to be similar when they share activities. At first, up to event window W^3, case c_4 seems to be quite similar to the other three cases seen so far. However, in window W^4 this changes due to activity f which has replaced activity e. This change is reflected in the similarity matrices M^1 to M^5. Consequently the value for Δ_3 indicates a possible change point in W^4.

5   Comparative Clustering

The previous two sections describe how to identify change in process behavior based on the change in similarity between cases. Once interesting change points have been identified, different clusterings can be created and compared, both programmatically and visually, in order to analyze the effect of changing behavior in a process. By comparing clusterings created for different sublogs (i.e. before and after an identified change point), we can see where behavior changed and analyze why. Of course, it is also possible to manually create selections of cases based on time or data attributes in order to compare behavior. This can, for example, be used to compare behavior for different age groups, for patients from different geographical locations, different years, etc.

Due to the properties of the MCL algorithm discussed in Sects. 3 and 4, changing behavior is reflected in changes in the cluster structure. For example, cluster sizes indicate the frequency of the captured behavior. Behavior that used to be common but is becoming less frequent will still be clustered separately rather than get merged with different or common behavior, as long as the behavior remains dissimilar.

Clusters of cases are represented as nodes in a partially connected, undirected graph. Node sizes indicate the number of clustered cases. Two nodes are connected when there is a positive similarity between at least one pair of cases between the two represented clusters. In more advanced visualizations, these edges could be given a weight to represent minimum, maximum, or average similarity, etc. This can be used to visually show inter-cluster similarity. Also, edge weights can be used in layout calculation for complex cluster graphs. Clusters are annotated with descriptions about the cases that are present. Shared activities between traces and similar data values are a few of the possibilities. For example, a cluster that groups cases of patients that all share a certain diagnosis can be annotated with that diagnosis description.

In Fig. 6, two example clusterings are compared. In the left clustering, one cluster is selected. By highlighting those clusters in the right clustering that share behavior with the selected cluster(s) on the left, we can interactively analyze how behavior has changed. It is possible to see how cases are clustered or what annotations are shared. For example, we might compare two sublogs of patient data, before and after an identified change in behavior. Patients that are present

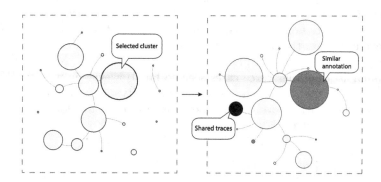

Fig. 6. Two clusterings are compared. Highlighted clusters on the right clustering show how behavior captured by the cluster selected on the left has changed.

in both years are highlighted as shared cases (in dark gray), while clusters that share some or all diagnoses (similar annotations) are highlighted (in light gray). Within a case clustering, clusters that share behavior are connected. As such, when comparing two clusterings, clusters that are split into or have emerged from multiple clusters can be found by looking at the clusters sharing annotations and/or cases. This can, for example, indicate that behavior has become more specific or more general. Besides visual approaches, techniques such as [20] that aim to explain clusters of cases can be used as well.

6   Evaluation

In order to evaluate our change detection technique we use a synthetic event log and two publicly available real-life event logs. In Subsect. 6.1 we show how change in behavior over time can be detected using our approach. In Subsect. 6.2 we apply our technique to uncover useful and interesting insights from real-life data. Our technique has been implemented in the process mining tool ProM[2], and is publicly available through the *TraceClustering* package.

6.1   Synthetic Evaluation

For the synthetic evaluation, an event log was generated from a fictive, manually created radiology process with 17 unique activities. 1,000 cases were generated spanning one year. The control-flow of this process heavily depends on the data-attributes 'age' and 'bodypart', as well as the time of year the patient arrives. As a result, there are many possible control-flow variants recorded in the event log.

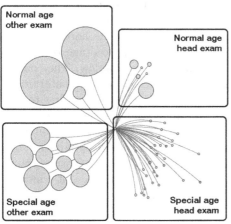

Fig. 7. Changes in behavior over time are hidden when clustering the entire synthetic event log and disregarding time.

We distinguish four different scenarios, based on patient age and the part of the body the radiology exam is to be made of. At the same time, the process has been constructed in such a way that temporal patterns in behavior occur, as the inflow of types of patients is seasonal. Special patients (patients younger than 10 years old that need a head exam) were modeled to arrive only in the months March and April, July and August, and in November and December. Patients of other ages and types can arrive all year long. In other words, the behavior of cases changes over time.

---

[2] See http://promtools.org.

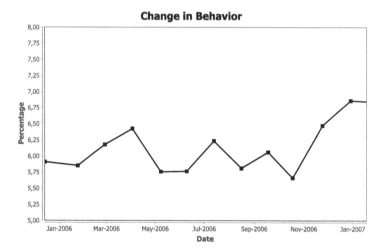

Fig. 8. Change in behavior over time for the synthetic event log. Changes in behavior can be identified by comparing case similarity matrices for consecutive time windows.

Figure 7 shows an example result of clustering the entire event log on the occurrence of activities, without regarding the time aspect. Clustering all cases in the log shows all behavior that was possible at any given point in time. As a result, we can only identify the different groups of patients, no changes in behavior or seasonal pattern are discovered.

By applying our change detection technique we can find changes in behavior over time. A new similarity matrix was created for the event log and compared with the previous matrix, every 100 events. Note that the choice of window size affects the perceived change, as larger window sizes smooth out local changes in behavior. Figure 8 shows the calculated change over time. The horizontal axis represents time whereas the vertical axis indicates the percentage of change occurring in each time window compared to the previous window. The change values indicate a change in behavior when the first special patients started to arrive in March. In Fig. 8, the age of cases is accounted for, according to the technique described in Subsect. 4.2. For every time window, the age factor of each case was multiplied by 0.95 in order to make sure that ages of older cases were accounted for. We can see that new change points are discovered in July and November, indicating a (potential) seasonal pattern, and interesting points in time to create new clusterings upon. As is expected, the first event windows indicate big change values since all information seen in them is new.

6.2   Real-Life Data

The first real-life log comes from a Dutch academic hospital that contains cases pertaining to cancer treatment procedures. It was originally used in the first Business Process Intelligence Contest (BPIC 2011) [15]. The second event log contains cases of building permit applications provided by a Dutch municipality.

This log is part of the 2015 edition of the BPI Challenge (BPIC 2015) [16]. These event logs were used so that the results can be reproduced. In the results shown here, we used the following MCL parameters: expansion = 2, inflation = 15.

The first event log contains cases of different stages of malignancy and of different parts of the body. Also, information is present about the diagnosis, treatment, specialism required, patient age, organisational group (hospital department), etc. This log contains 1,143 cases, 150,291 events and 624 distinct activities. There are 981 different executions paths (activity sequences). There are many attributes present on both the event and case level. All of these attributes can obtain several different values, leading to a large heterogeneity in the log. As cases are recorded between January 2005 and March 2008, the event log is likely to exhibit drifts in control-flow and changes in process behavior.

For each case in the hospital log, there are 16 attributes for 'diagnosis code', referring to the diagnoses the patient received for different parts of their body. By comparing on these attributes and calculating the change in behavior over time, we found that near July 2006, a change in diagnoses occurred. The change in behavior was calculated every 5,000 events, and every window the age factor of cases was multiplied by 0.95.

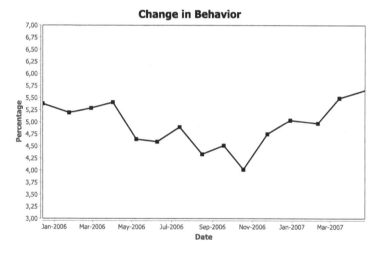

Fig. 9. Change in behavior over time for the hospital log, for the year 2006 and early 2007. Cases are compared on diagnosis code. Potential change in behavior is indicated in July 2006.

Figure 9 shows how the behavior has evolved. In Fig. 10, the clustering on the left represents cases two months before the change point whereas the clustering on the right represents cases two months after. Patients that were in the selected cluster and have had activities in both years are highlighted in dark gray. Groups of patients that have had (partially) shared diagnoses are marked light gray. We can see that some diagnoses are present for more body parts and now occur

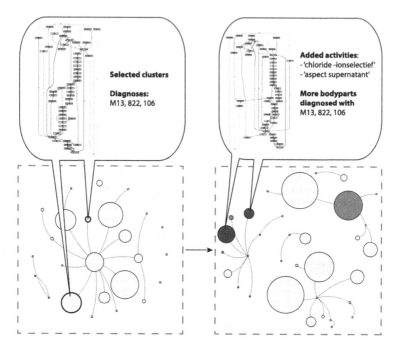

Fig. 10. Hospital log clustered on diagnosis code for cases active in May–June 2006 (left) and July–August 2006 (right). Changes in diagnoses are discovered. More body-parts are diagnosed with codes M13, 822 and 106.

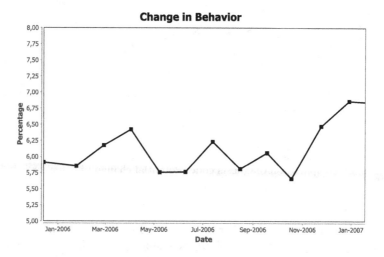

Fig. 11. Change in behavior over time for the hospital log compared on diagnosis code and treatment code, for the year 2006. Potential change in behavior is indicated in November 2006.

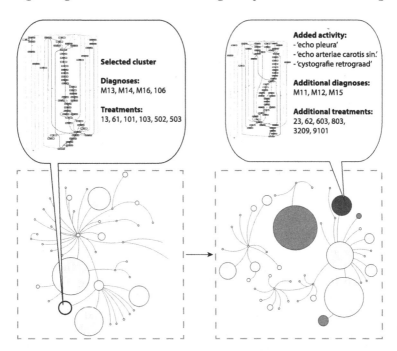

Fig. 12. Hospital log clustered on diagnosis code and treatment code for cases active in September–October 2006 (left) and November–December 2006 (right). Additional diagnoses and treatments are found.

in other combinations. This could indicate a trend in diseases or be due to an improvement in diagnosis detail. As there are many smaller clusters in July–August that have additional diagnoses (light gray), we can deduce that for the selected diagnosis, the related diagnoses have become more specific and diagnoses are also made on other parts of the body. In Fig. 10 process maps and differences in activities are shown for two highlighted clusters.

Besides diagnosis codes, every case has 16 possible attributes for 'treatment code', referring to the treatments the patient received on different parts of their body. This leads to many possible treatment combinations for different diagnoses. We inspect the change in behavior over time for the year 2006, when looking at diagnosis code and treatment code. As can be seen in Fig. 11, using our technique, a potential change point can be found in November 2006. By comparing the clustering results for September–October and November–December, changes in treatments for specific diagnoses become visible. As we can see from Fig. 12, treatment for some diagnoses have changed. Again cases active in the two months before the change point are shown on the left and cases active in the two months after are shown on the right. We can see that there are two clusters that contain patients that were active in September–October, one of which is much smaller than the other. For the larger cluster, additional diagnoses were made and additional treatments were performed. As a result, more cases share this behavior. The call-outs in Fig. 12 again show differences in activities between the two periods.

These differences indicate that, over time, treatments for certain diagnoses have changed. Considering the type of process, this could be due to specific patient needs, changes in protocols or advances in medicine. A probable reason is that diagnoses were made (or recorded) with greater detail. Insights such as these can be gained easily and can be used to verify or specify protocols, check whether certain behavior is changing or for auditing purposes.

The second real-life event log contains cases of building permit applications in a Dutch municipality. Information is present about the type of permit, the costs associated with the permit, the involved resources, etc. Again, each attribute can have several different values. This log contains 1,199 cases recorded between late 2010 and early 2015 with in total 52,217 events and 398 distinct activities. As there are 1,170 different execution paths, almost all cases are unique from the control-flow perspective.

Each case in the municipality log has an attribute 'parts' that refers to the different permit types that are involved in the case it describes. Each case is also labeled with the attribute 'term name', describing which status has been assigned to the permit application. Possible values are 'permit granted', 'additional information required', 'term objection and appeal', etc. Figure 13 shows the change in behavior over time when comparing cases on these two attributes, for the years 2012–2013. The change in behavior was calculated every 2,500 events, and in every window the age factor of cases was multiplied by 0.95. A potential change point is indicated near mid January 2013. We cluster the cases in the log on both permit type and term name and compare cases in December 2012–January 2013 with cases in February–March 2013. The results are shown in Fig. 14. As we can see, few clusters are discovered, indicating only slight differences in behavior on these perspectives. A group of cases pertaining to mainly construction and environmental permits that are in the 'objection and appeal' term is selected in the left clustering. During the selected period, most of this behavior has merged with the biggest group of cases, which now represents almost all behavior in the log in the right clustering.

6.3   Effect of Parameters

As explained, several parameters are important for obtaining the points in time where behavior has changed. Firstly, the perspectives used to create a case similarity matrix decide on which perspective change is detected. It is therefore important to choose those perspectives that are of interest to the analysis.

The MCL clustering technique uses two parameters, expansion and inflation, which both affect clustering granularity. When change on a low level is of interest, expansion can be decreased and inflation increased, and vice-versa for when only high-level change points are required. Besides the MCL parameters, the window size can also be adjusted to affect the detection span of the approach. Bigger window sizes will result in more global, high-level changes being detected while small windows will also reveal smaller changes in behavior.

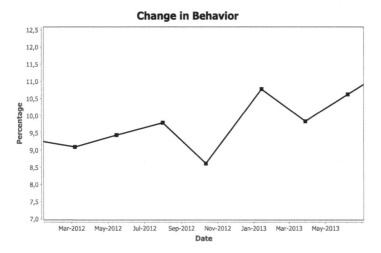

Fig. 13. Change in behavior over time for the municipality log compared on permit type and term description, for 2012–2013. Potential change in behavior is indicated in January 2013.

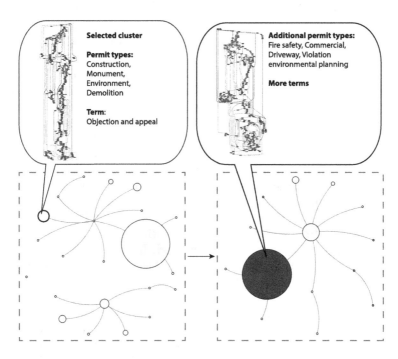

Fig. 14. Municipality log clustered on permit type and term description for cases active in December 2012–January 2013 (left) and February–March 2013 (right).

The effect of long-running cases and seasonal behavior can be controlled by adjusting the age factor. Increasing the age factor (to a value close to 1) will lead to longer lasting effects whereas decreasing the age factor will also show seasonal temporal behavior. In conclusion, the setting of parameter values needs to be decided on a case-per-case basis. Most interesting insights will be gained when the approach is used in an iterative process.

7   Related Work

Although concept drift is a well-studied topic in the data mining and machine learning communities, little work has been done on detecting concept drift in business processes. Bose et al. were the first to consider concept drift and change detection in a process mining setting [5]. In their work, a classification of possible changes in business processes is given, and statistical hypothesis tests are used to detect regions of change. Even tough the authors consider the possibility of change in data attributes, the scope of their work is limited to the detection of control-flow changes in a process manifested as sudden drifts over a period of time. More recently, Martjushev et al. built on this work by looking at gradual and multi-order dynamics to detect concept drift in control-flow [13]. They extend the work in [5] by providing solutions to detect gradual change as well. By considering multi-order dynamics through the use of an adaptive window technique, process change occurring at multiple levels of mixed time granularity can be detected. Maaradji et al. employ statistical tests over the distributions of runs observed in two consecutive time windows in order to detect concept drift [12]. As noted by the authors, in order to find differences in process behavior a notion of equivalence is necessary. In their paper, a notion of run-equivalence is used. It is shown that drift can be identified fast and accurately by using an adaptive sliding window technique. As a result, it can be used in an online (streaming) setting as an oracle as to when a discovered model should be updated.

Weber et al. employ probabilistic deterministic finite automata (PDFA) to represent the probability distributions generated by process models [19]. Similar to [5,12], statistical hypothesis tests are used to detect whether or not a distribution has changed significantly from a ground truth. The aim of their technique is to identify process change as soon as possible, but with confidence that change is significant, in order to discover a model representing reality as good as possible. As such, only drift in control-flow is considered. In [7] a different technique is proposed to automatically detect and manage concept drift in an online setting. Here, concept drift is detected real-time using an estimation technique based on abstract interpretation of the process and sequential sampling of the log. The fitness of prefixes of new samples taken from the log is checked against that of prefixes of initial samples. A change point is identified when there is a significant difference between these two points. In the above-mentioned techniques however, data attributes are not considered. As such, only changes in control-flow behavior can be discovered. Moreover, case maturity is not accounted for, leading to issues in discovering seasonal temporal changes in behavior.

Trace clustering techniques are often used to find different process variants. Several trace clustering techniques have been proposed in the field of process mining, and an extensive comparative analysis of trace clustering techniques has recently been performed in [14]. Often, however, the temporal dimension is not considered. In [11], the starting time of each process instance is used as an additional feature in trace clustering. By combining control-flow and time features, the clusters formed share both a structural similarity and a temporal proximity. The technique is based on the technique proposed in [4] and considers different types of changes, including sudden, recurring, gradual, and incremental changes. In more complex evolving business processes however, including the temporal proximity of cases might lead to misleading results. For example when seasonal drifts are intertwined with gradual changes in the process.

The technique proposed in this paper uses similar ideas and concepts as used in the papers mentioned above. However, trace clustering techniques and concepts are used to find changes in common and deviating process behavior. By taking into account both the control-flow and the data aspects, the technique is made context-aware. We extend the technique in [10] by including change detection in behavioral similarities between cases. The input is limited to the perspectives on which we want to cluster and compare behavior, and the two numerical parameters for the Markov cluster algorithm. It is not necessary to manually select the number of desired clusters, as that is determined by the underlying cluster algorithm, along with the cluster sizes and densities. Additionally, different windowing strategies, sizes and aging factors can be used to find different types of drift.

8   Conclusions and Future Work

Real-life *business processes* are often complex while exhibiting a high degree of variability. Due to changing conditions and circumstances, these processes continuously evolve over time. Existing *process mining* techniques assume the process to be static and are less suited for the analysis of contemporary business processes. In this paper we presented a novel *comparative case clustering* approach that is able to expose temporal changes in behavior in a process. By using both control-flow and case data we take the process context into account. Insights can be gained into how and why behavior has changed by comparing changes in clusterings over different partitions of the log. Interesting points in time can be discovered as to give an idea on where to partition the event log. The discovered information can then be used for further analysis, e.g. to design protocols, for early detection of unwanted behavior or for auditing purposes. Besides the time dimension, different data and control-flow attributes can be utilized in order to distinguish groups of behavior.

Our results show that indeed promising insights can be achieved. Nonetheless there are drawbacks. It is necessary to manually select the perspectives on which case similarity is calculated and what window size is used. Also, once change points have been identified, the parameters for the Markov cluster algorithm need to be chosen. Besides the parameters, at the moment, it is not possible to distinguish between changing behavior localized to a specific cluster and more global change. Additional research is needed to further automate the analysis process, for example by automatically detecting discriminating clustering perspectives or by suggesting parameters for the clustering algorithm. In the future we would also like to look into how changes in process behavior can be analyzed in an online, streaming event data, setting. Different ways to visualize change in behavior can be explored as well.

References

1. van der Aalst, W.M.P.: Process Mining: Discovery, Conformance and Enhancement of Business Processes. Springer, Heidelberg (2011)
2. van der Aalst, W.M.P., et al.: Process Mining Manifesto. In: Daniel, F., Barkaoui, K., Dustdar, S. (eds.) BPM 2011. LNBIP, vol. 99, pp. 169–194. Springer, Heidelberg (2012). doi:10.1007/978-3-642-28108-2_19
3. Bose, R.P.J.C., van der Aalst, W.M.P.: Context aware trace clustering: towards improving process mining results. In: Proceedings of the SIAM International Conference on Data Mining, pp. 401–412. Society for Industrial and Applied Mathematics (2009)
4. Bose, R.P.J.C., van der Aalst, W.M.P.: Trace clustering based on conserved patterns: towards achieving better process models. In: Rinderle-Ma, S., Sadiq, S., Leymann, F. (eds.) BPM 2009. LNBIP, vol. 43, pp. 170–181. Springer, Heidelberg (2010). doi:10.1007/978-3-642-12186-9_16
5. Bose, R.P.J.C., van der Aalst, W.M.P., Žliobaitė, I., Pechenizkiy, M.: Handling concept drift in process mining. In: Mouratidis, H., Rolland, C. (eds.) CAiSE 2011. LNCS, vol. 6741, pp. 391–405. Springer, Heidelberg (2011). doi:10.1007/978-3-642-21640-4_30
6. Burattin, A., Cimitile, M., Maggi, F.M., Sperduti, A.: Online discovery of declarative process models from event streams. IEEE Trans. Serv. Comput. 8(6), 833–846 (2015)
7. Carmona, J., Gavaldà, R.: Online techniques for dealing with concept drift in process mining. In: Hollmén, J., Klawonn, F., Tucker, A. (eds.) IDA 2012. LNCS, vol. 7619, pp. 90–102. Springer, Heidelberg (2012). doi:10.1007/978-3-642-34156-4_10
8. Gama, J.: Knowledge Discovery from Data Streams. CRC Press, Boca Raton (2010)
9. Goedertier, S., De Weerdt, J., Martens, D., Vanthienen, J., Baesens, B.: Process discovery in event logs: an application in the telecom industry. Appl. Soft Comput. 11(2), 1697–1710 (2011)
10. Hompes, B.F.A., Buijs, J.C.A.M., van der Aalst, W.M.P., Dixit, P.M., Buurman, J.: Discovering deviating cases and process variants using trace clustering. In: Proceedings of the 27th Benelux Conference on Artificial Intelligence (BNAIC), 5–6 November, Hasselt (2015)
11. Luengo, D., Sepúlveda, M.: Applying clustering in process mining to find different versions of a business process that changes over time. In: Daniel, F., Barkaoui, K., Dustdar, S. (eds.) BPM 2011. LNBIP, vol. 99, pp. 153–158. Springer, Heidelberg (2012). doi:10.1007/978-3-642-28108-2_15

12. Maaradji, A., Dumas, M., Rosa, M., Ostovar, A.: Fast and accurate business process drift detection. In: Motahari-Nezhad, H.R., Recker, J., Weidlich, M. (eds.) BPM 2015. LNCS, vol. 9253, pp. 406–422. Springer, Heidelberg (2015). doi:10. 1007/978-3-319-23063-4_27

13. Martjushev, J., Bose, R.P.J.C., van der Aalst, W.M.P.: Change point detection and dealing with gradual and multi-order dynamics in process mining. In: Matulevičius, R., Dumas, M. (eds.) BIR 2015. LNBIP, vol. 229, pp. 161–178. Springer, Heidelberg (2015). doi:10.1007/978-3-319-21915-8_11

14. Thaler, T., Ternis, S.F., Fettke, P., Loos, P.: A comparative analysis of process instance cluster techniques. In: Proceedings of the 12th International Conference on Wirtschaftsinformatik. Internationale Tagung Wirtschaftsinformatik (WI-15), 3–5 March, Osnabrck. Universitt Osnabrck (2015)

15. Dongen, B.F.: Real-life event logs - hospital log (2011). doi:10.4121/uuid: d9769f3d-0ab0-4fb8-803b-0d1120ffcf54

16. van Dongen, S.: BPI challenge 2015 (2015). doi: 10.4121/uuid:31a308ef-c844-48da-948c-305d167a0ec1

17. Van Dongen, S.: A cluster algorithm for graphs. Technical report, National Research Institute for Mathematics and Computer Science in the Netherlands (2000)

18. Veiga, G.M., Ferreira, D.R.: Understanding spaghetti models with sequence clustering for ProM. In: Rinderle-Ma, S., Sadiq, S., Leymann, F. (eds.) BPM 2009. LNBIP, vol. 43, pp. 92–103. Springer, Heidelberg (2010). doi:10.1007/978-3-642-12186-9_10

19. Weber, P., Bordbar, B., Tino, P.: Real-time detection of process change using process mining. In: Imperial College Computing Student, Workshop, pp. 108–114 (2011)

20. Weerdt, J., vanden Broucke, S.: SECPI: searching for explanations for clustered process instances. In: Sadiq, S., Soffer, P., Völzer, H. (eds.) BPM 2014. LNCS, vol. 8659, pp. 408–415. Springer, Heidelberg (2014). doi:10.1007/978-3-319-10172-9_29

Using Domain Knowledge to Enhance Process Mining Results

P.M. Dixit[1,2(✉)], J.C.A.M. Buijs[2], Wil M.P. van der Aalst[2], B.F.A. Hompes[1,2], and J. Buurman[1]

[1] Philips Research, Eindhoven, The Netherlands
{prabhakar.dixit,hans.buurman}@philips.com
[2] Department of Mathematics and Computer Science,
Eindhoven University of Technology, Eindhoven, The Netherlands
{j.c.a.m.buijs,b.f.a.hompes,w.m.p.v.d.aalst}@tue.nl

Abstract. Process discovery algorithms typically aim at discovering process models from event logs. Most algorithms achieve this by solely using an event log, without allowing the domain expert to influence the discovery in any way. However, the user may have certain domain expertise which should be exploited to create better process models. In this paper, we address this issue of incorporating domain knowledge to improve the discovered process model. First, we present a verification algorithm to verify the presence of certain constraints in a process model. Then, we present three modification algorithms to modify the process model. The outcome of our approach is a Pareto front of process models based on the constraints specified by the domain expert and common quality dimensions of process mining.

Keywords: User guided process discovery · Declare templates · Domain knowledge · Algorithm post processing

1 Introduction

Process mining aims to bridge the gap between big data analytics and traditional business process management. This field can primarily be categorized into (1) process discovery, (2) conformance checking and (3) enhancement [22]. Process discovery techniques focus on using the event data in order to discover process models. Conformance checking techniques focus on aligning the event data on a process model to verify how well the model fits the data and vice versa [2]. Whereas enhancement techniques use event data and process models to repair or enrich the process model. Process models can be represented by multiple modeling notations, for example BPMN, Petri nets, process trees, etc.

Most often, the focus of process discovery techniques is on automatically extracting information to discover models solely based on the event logs. Hence, in order to gain significant outcomes, the event log should ideally contain all the

P. Ceravolo and S. Rinderle-Ma (Eds.): SIMPDA 2015, LNBIP 244, pp. 76–104, 2017.
DOI: 10.1007/978-3-319-53435-0_4

necessary information required by the algorithm. However, in many real world scenarios, the data contained in the event logs may be incomplete and/or noisy. This could in turn result in incorrect interpretations of the data. One of the ways to circumvent the problems with insufficient data could be to allow the user to input certain expert knowledge. The domain knowledge could thus be used to overcome the shortcoming of the data; thereby improving the results. In this paper, we address this challenge of incorporating domain knowledge in traditional process discovery.

Domain knowledge can be introduced in the discovery process at multiple stages as shown in Fig. 1. In this paper, we focus on *post-processing* an already discovered process model to incorporate the user specified domain knowledge. The primary reason of introducing domain knowledge at the post-processing stage is to keep the overall method generic and scalable. Our approach can be coupled with *any* discovery algorithm which generates models which can be

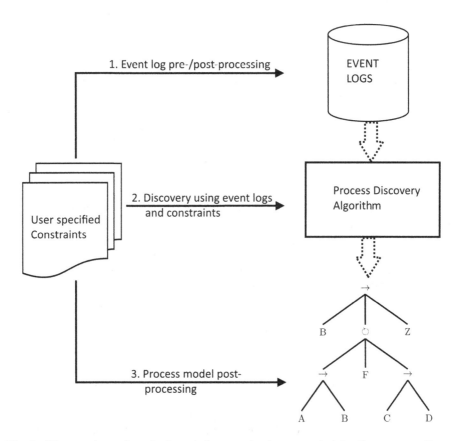

Fig. 1. Places where domain knowledge can be incorporated in the process discovery approach. In this paper we focus on post-processing the model based on domain knowledge (approach 3).

represented as process trees. Alternatively, we can refine (enhance) a pre-existing model using our technique. We focus on a class of models represented as process trees because the state of the art discovery algorithms such as the Inductive Miner [16] and the Evolutionary Tree Miner [5] discover block-structured process models represented by *process trees*. Furthermore, process trees are hierarchically structured and sound by construction. The hierarchical nature of process trees allows for a structured way to incorporate and validate the domain knowledge.

We use *Declare* templates [18] as a way to input the domain knowledge. *Declare* templates belong to the class of declarative languages, which are used to construct constraint based declarative process models. We abstract these templates as a way to specify domain knowledge effectively in terms of constraints. Declare provides a handy and effective way of modeling protocols and rules. Such constrains are especially efficient in many real life domains, such as healthcare wherein the the clinical specialist are usually aware of certain protocols which should hold in the process. Using a subset of declare templates the user can specify a set of constraints which must hold in the process.

In order to incorporate domain knowledge in the post-processing stage, we primarily focus on addressing two main challenges. Firstly, we introduce a verification algorithm, to determine whether a set of user specified constraints is satisfied in the model. We then introduce and evaluate three modification techniques in order to generate variants of the input process model based on the constraints and event log. In [10], we introduced the verification and a brute force modification approach. We extend that work and introduce two additional modification approaches: genetic and constraint specific modification. All three modification algorithms output a number of process tree variants. Each of these variants may satisfy the user specified constraint to a different degree. The user can choose the most appropriate model depending on the constraints satisfied, along with the values of four quality dimensions (replay fitness, precision, simplicity and generalization).

The remainder of the paper is structured as follows. In Sects. 2 and 3, we provide a literature review of related work and the preliminaries respectively. In Sects. 4 and 5 we explain the verification and modification algorithms. In Sect. 6 we evaluate our approach qualitatively as well as quantitatively based on synthetic and real life event logs. In Sect. 7 we conclude and discuss future research.

2   Related Work

Although the field of process discovery has matured in recent years, the aspect of applying user knowledge for discovering better process models is still in its nascent stages. Conformance techniques in process mining such as [1,2,8] replay event logs on the process model to check compliance, detect deviations and bottlenecks in the model. These techniques focus on verifying the conformance of event logs with a process model, but do not provide any way of incorporating domain knowledge to repair/improve the process model. [19] provides a backward compliance checking technique using alignments wherein the user provides compliance rules as Petri-nets. The focus is on using Petri-net rules for diagnostic analysis on the event log, rather than discovering a new process model with compliance rules. The conformance based repair technique suggested by [11] takes a process model and an event log as input, and outputs a repaired process model based on the event log. However, the input required for this approach is an end-to-end process model and a noise free event log. Our approach requires only parts of process models or constraints described using declarative templates.

[12] and [21] propose approaches to verify the presence/absence of constraints in the process model. However, these approaches do not provide a way to modify the process models w.r.t. the constraints. In [17], the authors provide a way to mine declarative rules and models in terms of LTL formulas based on event logs. Similarly, [7] uses declare taxonomy of constraints for defining artful processes expressed through regular expressions. [15] uses event logs to discover processes models using Inductive Logic Programming represented in terms of a sub-set of SCIFF [3] language, used to classify a trace as compliant or non compliant. In [6], the authors extend this work to express the discovered model in terms of Declare model. However, these approaches focus on discovering rules/constraints from event log without allowing the users to introduce domain knowledge during rule discovery.

In [20], authors suggest an approach to discover a control flow model based on event logs and prior knowledge specified in terms of augmented Information Control Nets (ICN). Our approach mainly differs in the aspect of gathering domain knowledge. Although declarative templates can also be used to construct a network of related activities (similar to ICN), it can also be used to provide a set of independent pairwise constraints or unary constraints. [23] proposes a discovery algorithm using Integer Linear Programming based on the theory of regions, which can be extended with a limited set of user specified constraints during process discovery. In [13], the authors introduce a process discovery technique presented as a multi-relational classification problem on event logs. Their approach is supplemented by Artificially Generated Negative Events(AGNEs), with a possibility to include prior knowledge (e.g. causal dependencies, parallelism) during discovery. The authors of [14] incorporate both positive and negative constraints during process discovery to discover C-net models. Compared to [13,14,23], our approach differs mainly in two aspects. Firstly, we do not propose a new process discovery algorithm, but provide a generic approach to post process an already discovered process tree. Secondly, our approach provides the user with a balanced set of process models which maximally satisfy user constraints and score high on quality dimensions.

3 Preliminaries

As mentioned in Sect. 1, we primarily use process trees to represent the process models and *Declare* templates as a means to incorporate the domain knowledge. This section provides a background and a brief description about process trees and *Declare* templates.

3.1   Process Trees

Process trees provide a way to represent process models in a hierarchically structured way containing operators (*parent nodes*) and activities (*leaf nodes*). The operator nodes specify control flow constructs in the process tree. Figure 2 shows an example process tree and its equivalent BPMN model. A process tree is traversed from left to right and top to bottom.

The order of child nodes is not important for *and* (∧), *exclusive-or* (×) and *inclusive-or* (∨) operators, unlike *sequence* (→) and *Xor-loop* (↻) where the order is significant. In the process tree from Fig. 2a, activities A and Z are always the first and last activities respectively. For the ↻ operator the left most node is the 'do' part of the loop and is executed at least once. In Fig. 2a, activity D is the optional 're-do' part of ↻, execution of which activates the loop again. Activities B and C occur in parallel and hence the order is not fixed. The right node of the loop is the escape node and it is executed exactly once. For the × operator, only one of either F or G is chosen. For the ∨ operator both × and activity E can occur in any order, or only one of either two can occur.

3.2   Declare Templates

A declarative model is defined by using constraints specified by a set of templates [18]. We use a subset of *Declare* templates as a way to input domain knowledge.

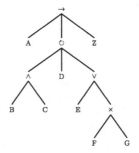

(a) Example Process tree showing *sequence* (→), *and* (∧), *exclusive-or* (×), *inclusive-or* (∨) and *Xor-loop* (↻) operators

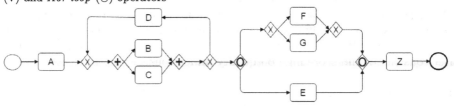

(b) BPMN equivalent of the process tree from Figure 2a

Fig. 2. Example process tree and its BPMN equivalent.

Table 1. Declare constraints and their graphical and textual interpretations

Template name	Graphical representation	Interpretation
response(A,B)		Activity B should (always) eventually occur after activity A
precedence(A,B)		Activity B can occur only after the occurrence of activity A
coexistence(A,B)		Activity A implies the presence of activity B (and vice versa)
responded-existence(A,B)		Activity B should (always) occur before or after the occurrence of activity A
not-coexistence(A,B)		Activity A implies the absence of activity B (and vice versa)
not-succession(A,B)		Activity A should never be followed by activity B
existence(n1,n2,A)		Activity A should occur: • n1..n2 times

Table 1 provides an overview and interpretation of the Declare constraints that we consider [17,18]. Binary templates provide ways to specify dependency (positive and negative) between two activities. For example, *response(A,B)* specifies that activity A has to be eventually followed by activity B somewhere in the process. We use six binary constraints as shown in Table 1. We use one unary constraint *existence(n1,n2,A)*, as a way to specify the range of occurrence of an activity.

4 Verification

In this section, we present a novel verification approach that takes a process tree and a set of constraints as input, and returns the set of constraints satisfied by the process tree as output. We recall from Sect. 3.1 that the order of children for the operators $\vee$, $\times$, and $\wedge$ is not fixed and the child nodes can be executed in any order. However, for the remaining nodes, $(\rightarrow, \circlearrowright)$ the order of navigation is fixed from left to right. We split the verification procedure into three parts. First, we check if the positioning of the activities in the process tree is correct according to the constraint. For example, for a constraint $response(A, B)$, we check if all the B's are on the *right* side of A. Next, we perform an additional check to verify the common parent, addressed in Subsect. 4.2. In the case of $response(A, B)$,

---

Algorithm 1. Declare constraints verification in a process tree

---

Input: process tree, set of constraints
Output: constraints categorized as *verified* or *unverified*
1 **begin**
2   **foreach** *constraint* **do**
3   **if** *not existence constraint* **then**
4   compute collection of *common* sub-trees
5   **foreach** *sub-tree* **do**
6   verify common parent
7   verify position of activties
8   **if** *common parent or position verification fails* **then**
9   set constraint verification unsuccessful
10   **else if** *relations constraint &occurs(ST_A,B) is (always)* **then**
11   set constraint verification successful
12   **else if** *negative relations constraint &occurs(ST_A,B) is (never)* **then**
13   set constraint verification successful
14   **else**
15   set constraint verification unsuccessful
16   **else**
17   consider full tree
18   check range from *occurs(PT,A)* to *occurs_multiple_times(PT,A)*
19   **return** *set of constraints marked as* - verified or unverified

---

even if activity B is on the *right* side of activity A, the order isn't fixed if the common parent between A and B is not → or ↻ and hence B is not guaranteed to occur after A. Finally, we calculate the occurrence possibilities of B w.r.t. the common parent. In Algorithm 1, we show the main sequence of steps used by the verification approach. In the following sub-sections, we detail the algorithm.

4.1 Sub-tree Computation and Position Verification

In this sub-section we explain the sub-tree computation and position verification with respect to each of the binary constraints. Sub-trees are sub-blocks containing the first common ancestor between the two activities of the binary (negative) relation constraints. The same activity can be present at multiple locations in a process tree which could result in multiple sub-trees for a single constraint. The total number of sub-trees is equal to the number of occurrences of the 'primary' activity from the constraint in the process tree. The primary activity is the activity w.r.t. which the constraint is specified. For e.g., for constraint *response(A,B)*, the primary activity is 'A'. For constraint such as *coexistence(A,B)*, where both the activities A and B are primary activities, the total number of sub-trees computed is $n + m$, where n and m are the number of occurrences of activities A and B in the process tree respectively. Table 2 provides the overview of primary activities corresponding to each constraint.

Table 2. Primary activities corresponding to binary constraints

Constraint	Primary activity
response(A,B)	*A*
precedence(A,B)	*B*
coexistence(A,B)	*A,B*
responded-existence(A,B)	*A*
not-succession(A,B)	*A*
not-coexistence(A,B)	*A,B*

For sub-tree calculation, we make use of the fact that a process tree is generally navigated from *left to right* for → and ↻, and that the ordering doesn't matter for other operators. This provides a good starting point to compare the ordering of activities in a process tree with respect to the declare constraints. Consider the constraint *response(A,B)* that should be verified for the process tree from Fig. 3a. As described in Table 1, a response constraint states that every occurrence of activity A should eventually be followed by activity B. In order to verify that such constraint holds true in the process tree, we first gather all the locations within the process tree where activity A occurs. For each occurrence of A in the process tree, we find the *first common ancestor* containing A and *all* the B's which can be reached after executing activity A. As discussed earlier, all the B's that could occur after A have to be on the *right* side of A. Figure 3b shows the sub-tree for constraint *response(A,B)*. Since there is only one occurrence of activity A in the process tree, there is only one sub-tree. The first occurrence

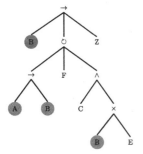

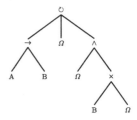

(a) Activities with cyan background are valid activities in process tree for constraint *response(A, B)*

(b) Sub-tree for *response(A, B)* with all the irrelevant activities marked as Ω

Fig. 3. Sub-tree computation for the constraint *response(A,B)*

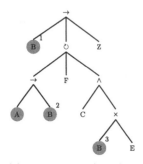

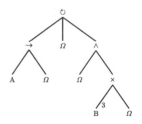

null

(b) *Precedence*(*A, B*) -
Null sub-tree with respect
to activity B_1

(c) *Precedence*(*A, B*) -
Sub-tree with respect to
activity B_2

(a) *Precedence*(*A, B*) -
Active nodes for this
constraint are highlighted
in cyan

(d) *Precedence*(*A, B*) -
Sub-tree with respect to
activity B_3

Fig. 4. Sub-trees computation for the constraint *precedence(A,B)*. As 4b results in a **null** sub-tree, the constraint verification for the constraint *precedence(A,B)* fails w.r.t. the entire process tree.

of B from the original process tree is ignored as it is on the *left* side of A, and hence this B cannot be guaranteed to be executed *after* executing activity A.

For the *precedence(A,B)* constraint; we are interested in finding all the common sub-trees with respect to B, containing all A's on the left side of (*executed before*) B. There are a total of 3 sub-trees corresponding to each B in the process tree from Fig. 4. The sub-trees for B_2 and B_3 are shown in Fig. 4c and d respectively. However, for B_1 there is no sub-tree containing activity A prior to (*i.e. on the left side of*) B. This results in a *null* sub-tree as shown in Fig. 4b, and therefore the verification fails.

Relation constraints such as coexistence and responded-existence are independent of the position of the other activity in the process tree. Figure 5 shows the sub-trees for constraints *responded − existence*(*A, B*) and *coexistence*(*A, B*). The sub-tree in Fig. 5d is calculated with respect to activity B and is only valid for the constraint *coexistence*(*A, B*). Negative relations constraints are more restrictive and the sub-trees can be calculated and derived in a similar way to their respective relation constraints counterpart. For example, for the constraint *not − coexistence*(*A, B*), the sub-trees can be computed similar to *coexistence*(*A, B*). However, unlike relation constraints, for negative relation constraints the absence of a sub-tree (*null* sub-tree) for each activity from constraint implies satisfaction of the constraint in the process tree. Sub-tree calculation is not necessary for unary constraints such as existence, wherein we consider the entire process tree. The next step is to determine if the common parent is valid as discussed in Subsect. 4.2.

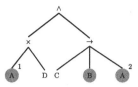

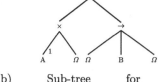

(a) *Responded − existence*(A, B) and *coexistence*(A, B) - Active nodes for these constraints are highlighted in cyan

(b)   Sub-tree   for   constraints   *coexistence*(A, B) and *responded − existence*(A, B) with respect to activity A_1

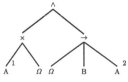

(c)   Sub-tree   for   constraints   *coexistence*(A, B) and *responded − existence*(A, B) with respect to activity A_2

(d)   Sub-tree   for   constraint *coexistence*(A, B) with   respect   to activity B

Fig. 5. Sub-trees computation for constraints *responded-existence(A,B)* and *coexistence(A,B)*

4.2 Parent Verification

If sub-tree computation is successful, then the next step is to verify the common parent. There are a set of *allowable* common parent operators for each type of constraint. For example, if we have to verify the *coexistence(E,B)* constraint on the process tree from Fig. 3a, then one of the sub-trees computed is shown in Fig. 6. As the common parent for this sub-tree is the choice operator ×, both E and B will never occur together. Hence the common parent verification for this particular sub-tree fails for constraint *coexistence(E,B)*. Parent verification is not required for unary constraints, as there is only one activity. Table 3 summarizes the valid common parents for all the binary constraints from Table 1.

Fig. 6. Sub-tree violating constraint *coexistence(E,B)*

4.3 Activity Occurrence Verification

For binary constraints the next step is checking the occurrence of the activity in the sub-tree. In order to achieve this, we use the predicate $occurs(ST_A, B)$, where A is the node with respect to which sub-tree ST is computed and B is the second activity of the binary constraint. For every ancestor of node A, we check the occurrence of activity B which can have the following values: *always, sometimes* or *never*.

Figure 7b shows the occurrence of activity B, for the sub-tree from Fig. 7a which is computed with respect to activity A. For choice operators such as ×

Table 3. Valid common parents for each of the declare constraints

constraint	valid common parent operator
response(A,B)	$\rightarrow, \circlearrowleft^1$
precedence(A,B)	$\rightarrow, \circlearrowleft^1$
coexistence(A,B)	$\rightarrow, \wedge, \circlearrowleft^1$
responded-existence(A,B)	$\rightarrow, \wedge, \circlearrowleft^1$
not-succession(A,B)	$\times$
not-coexistence(A,B)	$\times$
	$\circlearrowleft^1$ is valid only if node B (or A) is a child of the left (do) or right (exit loop) part and *not* of middle(re-do) part

and $\vee$, if activity B is present in *all* the child nodes, then activity B occurs *always* w.r.t. the operator node. If only few or none of the child nodes of the choice operator have occurrence of activity B, then activity B occurs *sometimes* or *never* resp. Similarly, if at least one child of $\rightarrow$ and $\wedge$ is activity B, then activity B occurs *always* w.r.t. this node. In case of $\circlearrowleft$ if activity B is present only in the re-do part of the loop (which may or may not be executed), then activity B occurs *sometimes*. If activity B is present in the loop or exit child of the $\circlearrowleft$ operator, then activity B is guaranteed to occur *always* w.r.t. this node. Starting with the deepest occurrence of B in sub-tree, we annotate the 'occurrence of B' w.r.t. each node recursively until the root of the sub-tree is reached as shown in Fig. 7b. We check the occurrence of activity B, at every ancestor of activity A. For binary relations constraint, if none of the ancestor(s) of activity A have the occurrence of B as *always*, then the constraint is not satisfied. On the contrary for negative relations constraints, if any of the ancestor(s) of activity A have the occurrence of B as *always* or *sometimes*, then the constraint is not satisfied.

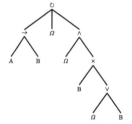

(a) Sub-tree for constraint *response(A, B)*

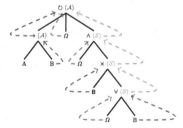

(b) Blue, red and green colors indicate the occurrence always($\mathcal{A}$), never($\mathcal{N}$) and sometimes($\mathcal{S}$) respectively.

Fig. 7. *Occurrence(ST$_A$, B)* verification for constraint *response(A,B)* (Color figure online)

In case of an unary constraint, the predicate *occurs_multiple_times(PT,A)* is calculated with possible values *yes* or *no*, where PT is the entire process tree and A is the activity from the unary constraint. If any of the ancestor(s) of activity A are children of the loop part or the re-do of ↻ operator, then the multiple occurrence of activity A is set to *yes*. Otherwise, the multiple occurrence part of activity A is set to *no*. *occurs_multiple_times(PT,A)* gives us the upper bound of the range, and we combine this with *occurs(PT,A)* to calculate the lower bound of the range. We evaluate the unary constraints at the root of the tree depending on the values of *occurs(PT,A)* and *occurs_multiple_times(PT,A)*, as shown in Table 4.

For binary constraints, if either the sub-tree computation, position verification, common parent verification or activity occurrence verification fails, then that constraint is marked as unsatisfied. If all these steps are successful for all the corresponding sub-trees, then the constraint is marked as satisfied. For unary constraints, if activity occurrence verification is successful (within the input range) then the constraint is marked satisfied, otherwise, it is marked as unsatisfied.

Table 4. Overview of possible ranges for existence constraint

occurs(PT,A) at the root of PT	*occurs_multiple_times(PT,A)* at the root of PT	range of occurrence
sometimes	*no*	0..1
sometimes	*yes*	0..n
always	*yes*	1..n
always	*no*	exactly 1
never	n.a	exactly 0

5   Modification

Following the description of the verification algorithm, we now discuss three modification approaches in order to incorporate the user specified domain knowledge in the process tree. Figure 8 gives a general overview of the overall approach, independent of the modification algorithm used. The three modification algorithms proposed are: brute force modification, genetic modification, and constraint specific modification. The brute force modification approach used only a process tree during modification. The constraint specific modification algorithm uses the user specified constraints along with the initial process tree during modification. The genetic modification approach requires the initial process tree, user specified constraints as well as the event log as during modification. It should be noted that in order to compute the final Pareto front, all the three techniques use the event log and the user specified constraints.

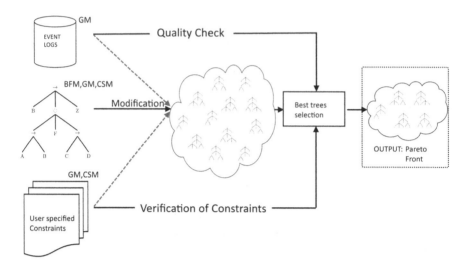

Fig. 8. The general overview combining traditional process discovery with domain knowledge specified using constraints. The figure also shows the inputs required by each modification approach - brute force modification(BFM), genetic modification(GM), and constraint specific modification(CSM). Each of the modification algorithms creates variants of process trees, after which the best process trees are selected using a Pareto front based on the four quality dimensions of process mining and number of constraints satisfied.

All modification algorithms produce a collection of candidate process trees as output. Each candidate process tree is evaluated against the four quality dimensions of process mining (replay fitness, precision, generalization and simplicity) [5,22] and the number of user specified constraints verified by the tree. This results in five quality dimensions. In order to evaluate the process trees based on these dimensions we use a Pareto front [5]. The general idea of a Pareto front is that all models are mutually non-dominating: A model is dominating with respect to another model, if for all measurement dimensions it is at least equal or better and for one strictly better. Using the five dimensions, a Pareto front is presented to the user which contains the set of dominating process trees.

5.1 Brute Force Modification

The brute force modification algorithm takes the discovered process tree as input and generates a list of candidate trees using a brute force approach. This is accomplished as shown in Algorithm 2, elaborated in the following steps:

1. Starting with the original input process tree, variants are created based on three primary edit operations: *Add node*, *Remove node* and *Modify node*.
2. Every node in the process tree is subject to each edit operation, resulting in new variants of process trees. After each edit operation, the number of edits for the process tree is incremented. It should be noted that, in case

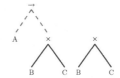

(a) Input Process tree. *(number of edits: 0)*

(b) Remove node: Removed activity A. Since $\rightarrow$ has only one child, it can be reduced. *(number of edits: 1)*

(c) Add node: Added activity A as a child of $\times$ *(number of edits: 2)*

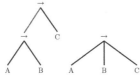

(d) Add node: Added operator $\rightarrow$ as a child of $\times$; and parent of A and B *(number of edits: 3)*

(e) Modify node: Modified operator $\times$ to $\rightarrow$; resulting in a process tree with only one parent operator *(number of edits: 4)*

(f) Remove node: Removed operator $\rightarrow$; resulting in an *empty* process tree *(number of edits: 5)*

Fig. 9. Example modification operations on process tree.

Algorithm 2. Brute force modification of a process tree

Input: process tree (PT), max nbr of edits(N_{max})
Output: candidate process trees collection(C)

1 **begin**
2   Create candidate process trees collection C
3     $C= \{pt\} \cup ModifyProcessTree(PT, 0, N_{max})$
4   **return** C;

5 **Function** *ModifyProcessTree(pt,n,N_{max})*
6   Initiate local process trees collection C'
7   **if** $n < N_{max}$ **then**
8   **foreach** node $\in$ process tree pt **do**
9   **foreach** edit operator **do**
10   create $\tilde{pt}$ from pt by using the edit operator for the chosen *node*
11   reduce tree $\tilde{pt}$ if possible
12              $C' = C' \cup \{\tilde{pt}\}$
13              $C' = C' \cup ModifyProcessTree(\tilde{pt}, n + 1, N_{max})$

14   **return** C'

---

of leaf nodes, *all* the activities are used exhaustively for *each* type of edit operation. For example, in Fig. 9c, activity A is added as a left-most child of $\times$, resulting in a new process tree. Similarly, activity B would also be added as a left-most child, resulting in another process tree and same goes for activity C. Moreover, each activity is also added to every possible location within the tree. For example in Fig. 9c, activity A could be added in between B and C resulting in a new process tree, as well as activity A would be added after C (*right of C*) resulting in a new process tree. Similarly, in case of operator nodes, *all* the other operator nodes are used exhaustively in each edit operation.

3. After every edit operation, the newly created tree is reduced recursively until no reductions are possible. The reduction primarily consists of the following two actions:
   - If an operator node contains only one child, then remove the operator node and replace it with its child node. (e.g. Fig. 9b)
   - If a parent and a child are of the same operator type in $\rightarrow, \wedge, \times$ or $\vee$; then the parent and child nodes can be collapsed into one operator. (e.g. Fig. 9e)
4. Each newly created variant of the process tree is further edited by iteratively calling all the edit operations exhaustively (in all possible orders) on each node using a "brute force" approach.
5. This process of creating process tree variants is repeated until all the nodes of all the process trees are processed and/or the threshold for the *number of edit* operations w.r.t. the process tree is reached.
6. Every variant of the process tree is added to the pool of candidate process trees. The candidate process trees might contain duplicates. The brute for approach could be computationally expensive, hence adding further processing to check for duplicates in a pairwise manner for all the candidate process trees would further degrade the performance. Therefore we do not remove the duplicates. However, the Pareto front presented to the user would take care of these duplicates from the candidate process trees.

Figure 9 shows different edit operations used by the modification algorithm. The *Modify node* operation modifies every node in the process tree and can be classified into *Modify activity* and *Modify operator* depending on the type of node selected. Similarly, *Add node* adds either an activity node or an operator node (Fig. 9c and d). An operator can be added below the parent node (Fig. 9d) and above the parent node (not shown in Fig. 9) by exhaustively combining child nodes. Each edit operation results in a new process tree, which can be further edited by other edit operations exhaustively until the threshold for edit distance is reached. Every process tree arising after each edit operation is added to the pool of candidate process trees. By executing all edit operations in an iterative way, we can find an optimal sequence of operations to deduce any process tree.

It is important to carefully set the threshold for maximum number of edit operations, as a high threshold could result in many changes and a small threshold would only explore a few changes in the resultant process tree as compared to the original process tree. Although a high threshold would explore more possibilities, it would result in a large number of variants of the process tree, and hence it would also be very computationally intensive and inefficient. Therefore, the threshold for number of edit operations should be chosen by the user depending on the original (starting) process tree, and the number of unverified constraints in the original process tree. That is, if the originally discovered process tree is completely inappropriate according to the domain expert, and the number of unverified constraints is high, then the threshold should be set high. Contrary to this, if the domain expert agrees with major elements of originally discovered process tree, and the number of unverified constraints is low, then the threshold should be set low.

5.2 Genetic Modification

The brute force approach from Subsect. 5.1 takes as input the process tree, and creates variants of this process tree in a brute force way, without considering the constraints or data. This approach takes into account the complete search space within the threshold of the number of edits. However, this could be unnecessarily time-consuming, as variants of process trees might be created iteratively which do not satisfy any constraints at all, or variants of process trees which do not describe the event log very well, or both. In order to overcome this, we present a second modification algorithm using a genetic approach. The creation of process trees is guided by the event log using the four quality dimensions (replay fitness, precision, generalization and simplicity) along with the number of constraints verified according to the verification algorithm.

We extend the genetic approach for process trees discussed in [4] to include an additional fitness evaluator "number of constraints specified", along with the four standard quality dimensions. A number of factors such as crossover and mutation probabilities, number of random trees in each generation etc. determine the variation of process trees in each generation. The tree modification process is guided to balance between the standard quality dimensions (determined by the event log) and the number of constraints verified. The modification of the trees is stopped when the stopping criteria of the genetic algorithm are met. The stopping criteria could be the maximum number of generations, maximum time taken, minimum number of constraints satisfied etc. The end result is a Pareto front containing the best process trees balanced on the five dimensions.

5.3 Constraint Specific Modification

The modification approaches discussed in Subsects. 5.1 and 5.2 first modify the process trees (either in a brute force way or genetically), and then evaluate them to check the compliance of user specified constraints on the modified trees. An alternative to this approach would be to change the process trees according to the type of constraint, rather than changing the trees first and then verifying. In order to achieve this, we introduce a constraint specific modification approach which iteratively changes the process trees for each constraint specified by the user. The resulting process trees are subject to modification with respect to the next constraint considering all possible permutations. The current implementation of constraint specific modification requires the input process tree to have non-duplicate labels. In future we aim to overcome this issue of duplication by taking in account all the combinations of each activity occurrence in the process tree, depending on the type of constraint. However due to the complexity and search space, we limit the current implementation to non-duplicate labels. The steps followed for constraint specific modification are mentioned below:

1. As a first step, all the possible permutations of the constraints given by the user are computed. For example, if there are two constraints input by the user - *response(A,B)*, *precedence(C,D)* then the set of possible permutations of these constraints is - (*<response(A,B)*, *precedence(C,D)>*; *<precedence(C,D)*, *response(A,B)>*)
2. For each permutation, the process tree is modified with respect to each constraint in a sequential order as shown in Fig. 10. Hence, new variants of process trees, are created and added to the output process trees pool. Each new variant is again subject to further modification, based on the next constraint in the set. The modification of process tree(s) based on the type of constraint is performed using following step(s):
   (a) For unary constraints, the activity from the constraint is mapped to the activity in the process tree. As discussed previously, we assume that every activity in a process tree is present at most once. Next, according to Table 4 if *occurs(PT,A)* of the activity is *sometimes*, then we introduce a choice operator as one of the ancestor of activity A (if it doesn't already exist at any ancestors). Each possible ancestor is changed, resulting in

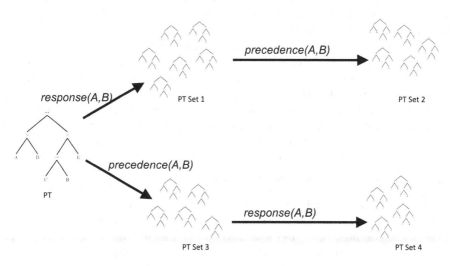

Fig. 10. High level representation of constraint specific modification approach. Starting with initial process tree, multiple variants are created w.r.t. each constraint. Every edited process tree is subject to the next constraint, until all the constraints are considered. This process is repeated for all the possible combinations of constraints. All the process trees in each PT Set are added to the final output of candidate trees for evaluation in the Pareto front.

creation of new process trees. Each newly created process tree is added to the pool of candidate process trees. Similarly, if *occurs(PT,A)* of the activity is *always*, then we replace every choice operator (one by one) with allowable operators ($\rightarrow,\wedge,\circlearrowleft$). Each resulting process tree is added to the pool of candidate process trees. Similar steps are repeated depending on the values of *occurs_multiple_times(PT,A)* and the presence/absence of loop operator in the ancestors of activity A. In case of binary constraints, we follow steps (b) to (e).

(b) For binary constraints, two activities from the constraint are used to create a process tree as shown in Fig. 11a. The root shows the allowable operators as common ancestor. The operators on the edge show the allowable intermediate operators between the root and node B.

(c) The common ancestor between these activities is checked. If the common ancestor between these activities is valid (as described in Table 3), then proceed to next the step. If the common ancestor of is not valid, then it is changed to an allowable valid operator from Table 3. This is done for every allowable valid operator and results in a set of process trees. Each process tree created is subject to the next steps.

(d) The next step is to check the intermediate operators between common parent and the secondary activity from the constraint in the process tree as shown in Fig. 11b. Constraint such as *co-existence(A,B)* is decomposed into two constraints *responded-existence(A,B)* and *responded-existence(B,A)*, which essentially add up to *co-existence(A,B)*. The allowable valid operators between common ancestor and secondary activity are mentioned in Table 5.

(e) In case the intermediate operator is not valid, it is replaced by each valid intermediate operator from Table 5. Every replacement of the operator results in a new process tree, which is added to the list of candidate process trees. Figure 11 shows the constraint specific modification process for the constraint *response(A,B)*. Figure 11c shows *one* of the outcomes of the constraint specification modification. It should be noted that some constraints share similar properties.

Table 5. Valid intermediate operators for each of the declare constraints

Constraint	Valid common parent operator
response(A,B)	$\rightarrow$, $\wedge$, $\circlearrowleft^1$
precedence(A,B)	$\rightarrow$, $\wedge$, $\circlearrowleft^1$
responded-existence(A,B)	$\rightarrow$, $\wedge$, $\circlearrowleft^1$
not-succession(A,B)	$\times$, $\rightarrow$, $\wedge$, $\circlearrowleft$, $\vee$ *(all operators)*
not-coexistence(A,B)	$\times$, $\rightarrow$, $\wedge$, $\circlearrowleft$, $\vee$ *(all operators)*
	$\circlearrowleft^1$ is valid only if node B (or A) is a child of the left (do) or right (exit loop) part and *not* of the middle(re-do) part

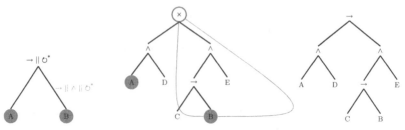

(a) Process tree showing allowable operators for constraint *response(A,B)*.

(b) Process tree highlighted with red as the common parent operator. The red area shows the intermediate operators between common parent and activity B.

(c) Modified process tree, satisfying the constraint *response(A,B)*, formed by combining process tree from Figure 11a and Figure 11b (by replacing × with →).

Fig. 11. Constraint specific modification for *response(A,B)* - *depicted as a process tree in* a on the process tree from b

3. Finally, the best variants are chosen using a Pareto front made of the number of constraints satisfied and the four standard quality dimensions of process mining. It should be noted that some of the constraints might share similar properties or might be related. In such cases, for certain combinations, modification of a tree to satisfy a certain constraint might inherently also lead to the satisfaction of another constraint. In such scenarios, there is no further modification required, if there are no unsatisfied constraints remaining for the current combination set. For example, the valid common operators as shown in Table 3 for $response(A, B)$ and $precedence(A, B)$ are same. Therefore if a process tree contains a single occurrence of activities A and B, then the modification to change the common parent for the constraint $response(A, B)$, would automatically contain the modifications for the constraint $precedence(A, B)$ too. Hence step (c) could be skipped when the modifications for $prcedence(A, B)$ are being performed.

6   Evaluation

The outcomes of the three modification approaches can be evaluated primarily in two ways. One way is to qualitatively evaluate the differences in each modification approach in terms of configuration, performance and output. The second approach is to quantitatively evaluate the outcomes of the three modification approaches, and to specifically determine the added benefit that domain knowledge brings in improving the process models. We first compare and discuss the peculiarities of the three modification approaches qualitatively based on different parameters. Next, we evaluate the outcomes of different modification approaches quantitatively using a synthetic and a real life event log.

6.1   Comparison of Approaches

Multiple parameters determine the outcome of each technique. Every algorithm has various pro's and con's associated with it. That is, one technique might excel in a certain scenario where another technique might fail and vice versa. Therefore it is important to first carefully evaluate each approach qualitatively to analyze and compare the properties associated with each. In this sub section we focus on the qualitative comparison of the three modification algorithms introduced in this paper.

Configuration. This section addresses the inputs and the configurable input parameters required by the modification techniques, which have a direct or indirect impact on the outcome of the algorithm. As discussed previously, all the three techniques require an event log, user specified constraints and a process tree as input. The constraint-specific modification technique does not require any additional inputs. The brute force modification technique requires the user to set the 'threshold for # of edits'. This threshold sets the maximum space explored by the algorithm, and hence has a big impact on the number of trees discovered. The genetic approach requires additional parameters which are standard for a genetic algorithm such as - maximum population, elite size, mutation probabilities etc. It should also be noted that the genetic algorithm is probabilistic whereas brute force and constraint specific modification techniques are deterministic. Table 6 gives an overview of the inputs required by the three techniques.

Table 6. Input and input parameters affecting the discovery outcome for each modification approach.

Constraint specific	Brute force	Evolutionary tree miner
Event log	Event log	Event log
Constraints	Constraints	Constraints
Process tree	Process tree	Process tree
	Threshold for # of edits	Population size
		Elite count
		Max generations
		Max duration
		Cross over probability
		Mutation probability

Event Log Usage. The way in which the information from event logs is used by the three approaches varies. Table 7 shows the usage of event logs in each approach. The brute force and the constraint specific modification techniques first create a set of candidate process trees without using any information from the event log. The event log is only used at a later stage to compute the quality scores with respect to each process tree variant, which is in turn used to construct a Parteo front. The genetic algorithm on the other hand uses the event logs along with the number of constraints verified as the guiding rule. Each process tree variant created is subject to log alignments. The majority of the trees which make it to the next generation score well in some (or all) of the quality dimensions.

Table 7. Usage of event logs of different algorithms.

Event log used during	Constraint specific	Brute force	Genetic
Final evaluation	✓	✓	✓
Intermediate evaluation	×	×	✓

Time Performance. The performance time of each algorithm varies depending on the size of input process tree, size of the event logs and most importantly, the individual algorithm settings. All the three modification approaches are subject to constraint verification and event log alignments as final evaluation to calculate the quality of candidate process trees. It should be noted that in the case of genetic modification there are also intermediate event log alignments with respect to every intermediate process tree. The genetic algorithm additionally depends on the setting of various parameters. For example, if the parameters such as population size and maximum generations are kept high, the genetic algorithm may take a long time to complete. The time complexity in the case of brute force modification technique depends on the 'threshold for maximum number of edits' and the number of nodes in the initial tree. However, the time complexity for constraint specific modification approach is determined by the number of constraints, the type of constraints and the depth of the initial tree. That is, if the number of constraints are high, the constraints are unrelated and the size of the initial tree is big then the constraint specific algorithm can take a long time to complete. In a real life setting, the constraint specific modification is usually faster than the genetic approach, which in turn is faster than brute force modification. This can be explained by the fact that constraint specific modification is primarily dependent on the number of constraints, hence is fastest. The genetic approach discards the uninteresting candidates in each generation, thereby converging faster than the brute force approach, which explores all the variants within the given threshold. Of course, as mentioned previously the individual parameters have a much higher impact on the performance time of each of the approaches.

Next, we focus on discussing the quantitative evaluation of the candidate process trees from the Pareto front. One method to evaluate the quality of outcome from each modification approach could be to present the domain expert with a list of candidate process trees (or process models) to choose from. The domain expert can choose the most appropriate process trees guided by the quality scores for each process tree. For e.g., if the domain expert wants to focus on process trees with high fitness, then Pareto front could be navigated to only look at process trees with high fitness. However this approach is highly subjective and would depend entirely on the preference of the domain expert, and hence would be difficult to conduct in a scientific manner. Another approach for evaluation is to discover an *expected* model based on user specified constraints. In this approach there is a certain expected model, which isn't discovered by the traditional process discovery techniques due to reasons such as data inconsistencies, discovery algorithm biases etc. We use the latter approach for evaluation as it provides a ground truth that can be used to quantify results, and be used to evaluate the results in a controlled way without depending on the high subjectivity of the domain expert. We evaluate our approach based on both a synthetic log and a real life log. The synthetic log demonstrates the usage of our approach on small process trees. The real life event log is from a road traffic fine management process.

6.2 Synthetic Event Log

We use a synthetic event log to demonstrate how our approach could improve an incorrect model discovered due to algorithm bias and noisy event log. For the event log $L = [\langle A,B,C,D \rangle^{90}, \langle A,C,B,D \rangle^{90}, \langle A,C,D,B \rangle^{90}, \langle C,A,D,B \rangle^{90}, \langle C,A,B,D \rangle^{90}, \langle C,D,A,B \rangle^{90}, \langle C,D,B,A \rangle^{6}, \langle C,B,A,D \rangle^{6}, \langle D,A,C,B \rangle^{6}]$, the Inductive Miner infrequent (IMi) [16] with default settings generates the process tree with all four activities in parallel as shown in Fig. 12a.

(a) Original process tree as discovered by Inductive Miner infrequent.   (b) Modified process tree present in the output of all modification approaches.

(c) Modified process tree present in the output of all modification approaches.   (d) Modified process tree present in the output of all modification approaches.

Fig. 12. Original and modified process trees for event log L.

Table 8. Quality dimensions of the Pareto front for process trees from Fig. 12

Tree	Constraints satisfied	Replay fitness	Precision	Generalization	Simplicity
Fig. 12a	0	1	0.833	0.957	1
Fig. 12b	4	0.997	1	0.957	1
Fig. 12c	2	0.999	0.900	0.957	1
Fig. 12d	2	0.998	0.993	0.957	1

Table 9. Performance statistics for the modification approaches from Sect. 5

Approach	Total # of trees discovered	Total # of trees in Pareto front	Average computation time for 5 runs (in ms)
Brute force	8323	88	53232
Genetic	50000[a]	112[b]	134689
Constraint specific	481	4	295

[a] *50000 is the maximum # of trees that could be produced (including duplicates).*
[b] *112 is the average number of trees in Pareto front over five runs.*

From the high frequent traces of the log we can deduce simple rules such as activity A is always eventually followed by activity B; and activity B is always preceded by activity A. Similar relationship holds for activities C and D. We use this information to deduce the following four constraints: *response(A,B)*, *precedence(A,B)*, *response(C,D)*, and *precedence(C,D)*. We use these constraints, the process tree from Fig. 12a discovered using IMi [16] and the synthetic log (L) as our input and perform the experiment with all three modification approaches discussed in Sect. 5. For the genetic modification approach, we set the stopping condition of maximum generations to 500 generations. The population size in each generation is set to 100. Along with this, we add some additional conditions restricting duplicate labels and guaranteeing minimum number of nodes. All the other settings are kept as default as in the ETM miner of [4]. In case of brute force modification, the maximum number of edits is set to 3.

Table 9 summarizes the performance statistics and output of each modification algorithm. The constraint specific modification algorithm outperforms the other approaches in terms of time taken, but also produces fewer process trees compared to the other two approaches. The Genetic algorithm produces a Pareto front containing varying number of process trees in each new run. Also, it takes the longest time as every process tree created in each generation is subject to alignments, which is often time consuming. The brute force modification is faster than the genetic modification as the number of nodes in the initial tree and the number of activities is very low. Figure 12 shows the original process tree discovered by IMi (Fig. 12a) and 3 modified process trees present in the Pareto front output of each modification approach. Table 8 summarizes the dimension scores of the process trees from Fig. 12.

The modified process tree from Fig. 12b satisfies all the four constraints. The tree from Fig. 12b also has a higher precision value of 1, and a considerably high replay fitness score. This process tree is highly precise, thereby explaining the high frequent traces of the event log much better and ignoring the infrequent noisy traces. Hence, if the user is interested the most in precision, then this is the process tree that would be chosen. However, if the user is a little relaxed w.r.t. to precision then process trees from Fig. 12c and d might be interesting. These process trees have a very high precision score along with very high fitness scores, outperforming the originally discovered process tree. However, only half the constrains specified are present in these process trees. Hence, if the user decides that some constraint(s) can be relaxed/ignored, then these process trees could be chosen by the user. Based on user's preferences, the Pareto front could be navigated to select the most interesting process models.

6.3   Real Life Event Log

Exceptional cases may dominate the normal cases, thereby leading to a process model that is over-fitting the data or that is too general to be of any value. This process model could however be improved by incorporating domain knowledge. In order to evaluate such a scenario, we use the following steps on a real-life log containing the road traffic fine management process with 11 activities and 150,370 cases available at [9]:

1. Use the complete event log to mine a process model using the heuristics miner. Learn *domain rules* based on this model and the log visualizers from ProM.
2. Filter the event log to select 1% of the cases having *exceptionally deviating* behavior.
3. Create a process tree based on the filtered log using inductive miner infrequent [16]. The BPMN representation of this process tree is shown in Fig. 13.
4. Use the domain rules learned in step 1, the process tree from the filtered log of step 3, in combination with the *complete* event log as input to each of the modification approaches.

We deduce 1 precedence, 1 response, 1 responded-existence and 1 not-succesion rule from the complete event log using the heuristics miner and the log visualizers from ProM. These rules are shown in Table 10. It should be noted that these rules may or may not hold in the *actual* real life process model corresponding to the event log. However, all these rules have a high support and confidence according to the event log, and hence we consider them to be true.

For the genetic modification approach, we set the stopping condition of maximum generations to 500 generations with the population size set to 100. All the other settings are kept as default from the ETM miner of [4]. Therefore, in total a maximum of 50000 trees are evaluated. In case of brute force modification, the maximum number of edits is set to 3. Furthermore, due to limited computing resources, the maximum number of trees selected for Pareto front evaluation in both brute force modification and constraint specific modification approach was

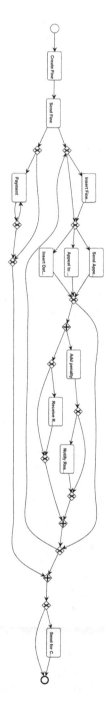

Fig. 13. BPMN model mined using IMi with filtered log containing infrequent traces only.

Table 10. Domain rules derived from the complete real life event log.

Activity 1	Constraint	Activity 2
Insert Fine Notification	Precedence	**Send Appeal to Perfecture**
Receive Result Appeal from Perfecture	Responded Existence	Notify Result Appeal to Offender
Appeal to Judge	Not Succession	Send Appeal to Perfecture
Insert Date Appeal to Perfecture	Response	Send Appeal to Perfecture

Activities in **bold** are the primary activities.

limited to first 5000 trees satisfying at least 50% of the constraints. That is, for constraint specific modification approach and brute force modification approach, the total number of trees kept in memory were 5000 each, which were then used to populate the Pareto front.

Table 11. Dimensions statistics for process trees based on real life event log

Tree	Constraints satisfied	Replay fitness	Precision	Generalization	Simplicity
IMi	0	0.74	0.52	0.84	1
Alpha	0	0.74	0.72	0.96	1
ILP	0	1	0.37	0.99	1
Constraint specific	4	0.99	0.64	0.99	1
Genetic	4	0.82	0.71	0.98	1
Brute force	4	0.83	0.68	0.99	1

In Table 11 we compare the evaluation outcomes of different approaches in terms of fitness, precision, generalization and simplicity. Since we considered the original complete event log as the ground truth, in the discussion to follow we primarily focus on the replay fitness and precision dimensions, as these two dimensions give a very good description of the model in terms of the event log. When the process model obtained from filtered event log (Fig. 13), is evaluated against the complete event log, it features very poorly with very low fitness and precision scores of 0.74 and 0.52 respectively. Furthermore, we used a few of the automated process discovery algorithms to discover a model based on the filtered event log. Each of these models were then evaluated against the complete event log, and the results presented in the Table 11. Model mined by ILP miner returned a perfect fitness score of 1, even against the complete event log. However, many of the activities in the discovered model were completely isolated. Hence, it allowed for any behavior thereby fitting the complete log very well, but having extremely poor precision value of 0.34. The alpha algorithm

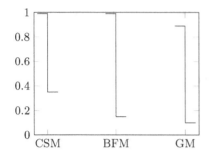
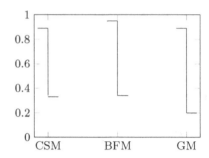

(a) Range of fitness scores of all the trees from the Pareto front for each algorithm.

(b) Range of precision scores of all the trees from the Pareto front for each algorithm.

Fig. 14. Range of fitness and precision scores for process trees from Pareto front output based on real life event logs.

generated a model which had the same score (of 0.74) for replay fitness as the IMi, and a much better precision value of 0.72.

As a next step we compare the outcomes of models mined using domain knowledge against the models mined without domain knowledge. For this we need to choose some models from the outputs of each of the modification approaches. All the three approaches contained at least 50 process trees in the Pareto fronts. As evident from Fig. 14, the range of fitness and precision values is big, for process trees from the Pareto front. In order to compare the outcome of the modification approaches, we selected three process trees, one corresponding to each modification approach having a balanced high score of fitness and precision, and satisfying all the four constraints. All the three models chosen have a considerably higher fitness scored than the traditional techniques, except the model mined by ILP miner. Similarly, the precision scores of each of the selected model were usually considerably higher than the traditional process automated discovery techniques. This makes the improvements after adding the domain knowledge for post processing quite evident.

7   Conclusions and Future Work

In this paper we introduced a way to incorporate the domain knowledge of the expert in an already discovered process model. We primarily introduced two types of algorithms: verification and modification, in order to verify and incorporate domain knowledge in the process model respectively. The proposed verification algorithm provides a comprehensive way of validating whether the constraints are satisfied by the process tree. In the current approach we consider a subset of *Declare* templates. In the future this could be extended to include all the *Declare* templates. For modification, we have proposed three algorithms. The brute force approach exhaustively generates multiple process trees. However, the brute force approach does not consider the user constraints during the modification process. In order to overcome this, we introduce a genetic approach to consider both log and constraints during modification. As shown in Subsect. 6.3, the guided genetic algorithm is considerably faster than the brute force approach when the number of activity classes is high. However, individually both brute force and genetic approach are time and memory inefficient. Hence, we introduced another modification approach which changes the process model by individually considering each user constraint. Currently, underlying assumption of this approach is that the process model does not contain duplicate labels. Each modification approach has its own set of pro's and con's. For example, brute force modification and constraint specific modification techniques result in a consistent Pareto front, on repetitive runs of experiments. However, in the case of genetic approach, different runs could result in different process trees in the Pareto front. The constraint specific modification approach performs a pre-existing set of operations for modification and does not consider log information while modification, hence is faster than other approaches but rather limited in the number of process trees created. In the future, we would like to optimise the modification approach and/or ensure certain guarantees in the modified process trees. Another future direction could be to incorporate domain knowledge at different stages, for example when logging event data or during the discovery phase.

References

1. Adriansyah, A., van Dongen, B.F., van der Aalst, W.M.P.: Conformance checking using cost-based fitness analysis. In: 15th IEEE International Enterprise Distributed Object Computing Conference (EDOC), pp. 55–64. IEEE (2011)
2. Adriansyah, A., Dongen, B.F., van der Aalst, W.M.P.: Towards robust conformance checking. In: Muehlen, M., Su, J. (eds.) BPM 2010. LNBIP, vol. 66, pp. 122–133. Springer, Heidelberg (2011). doi:10.1007/978-3-642-20511-8_11
3. Alberti, M., Chesani, F., Gavanelli, M., Lamma, E., Mello, P., Torroni, P. Verifiable agent interaction in abductive logic programming: The sciff framework. ACM Trans. Comput. Logic 9(4), 29:1–29:43 (2008)
4. Buijs, J.C.A.M., van Dongen, B.F., van der Aalst, W.M.P.: A genetic algorithm for discovering process trees. In: IEEE Congress on Evolutionary Computation (CEC), pp. 1–8. IEEE (2012)
5. Buijs, J., van Dongen, B.F., van der Aalst, W.M.P. Quality dimensions in process discovery: the importance of fitness, precision, generalization and simplicity. Int. J. Cooperative Inf. Syst., 23(1) (2014). doi:10.1142/S0218843014400012
6. Chesani, F., Lamma, E., Mello, P., Montali, M., Riguzzi, F., Storari, S.: Exploiting inductive logic programming techniques for declarative process mining. In: Jensen, K., van der Aalst, W.M.P. (eds.) ToPNoC II. LNCS, vol. 5460, pp. 278–295. Springer, Heidelberg (2009). doi:10.1007/978-3-642-00899-3_16
7. Ciccio, C., Mecella, M.: Mining constraints for artful processes. In: Abramowicz, W., Kriksciuniene, D., Sakalauskas, V. (eds.) BIS 2012. LNBIP, vol. 117, pp. 11–23. Springer, Heidelberg (2012). doi:10.1007/978-3-642-30359-3_2

8. Leoni, M., Maggi, F.M., van der Aalst, W.M.P.: Aligning event logs and declarative process models for conformance checking. In: Barros, A., Gal, A., Kindler, E. (eds.) BPM 2012. LNCS, vol. 7481, pp. 82–97. Springer, Heidelberg (2012). doi:10.1007/978-3-642-32885-5_6

9. de Leoni, M., Mannhardt, F.: Road traffic fine management process. http://dx.doi.org/10.4121/uuid:270fd440-1057-4fb9-89a9-b699b47990f5

10. Dixit, P.M., Buijs, J.C.A.M., van der Aalst, W.M.P., Hompes, B.F.A., Buurman, J.: Enhancing process mining results using domain knowledge

11. Fahland, D., van der Aalst, W.M.P.: Repairing process models to reflect reality. In: Barros, A., Gal, A., Kindler, E. (eds.) BPM 2012. LNCS, vol. 7481, pp. 229–245. Springer, Heidelberg (2012). doi:10.1007/978-3-642-32885-5_19

12. Giordano, L., Martelli, A., Spiotta, M., Dupre, D.T.: Business process verification with constraint temporal answer set programming. Theory, Pract. Logic Program. **13**, 641–655 (2013). doi:10.1017/S1471068413000409

13. Goedertier, S., Martens, D., Vanthienen, J., Baesens, B.: Robust process discovery with artificial negative events. J. Mach. Learn. Res. **10**, 1305–1340 (2009)

14. Greco, G., Guzzo, A., Lupa, F., Luigi, P.: Process discovery under precedence constraints. ACM Trans. Knowl. Discov. Data **9**(4), 2:1–32:39 (2015)

15. Lamma, E., Mello, P., Riguzzi, F., Storari, S.: Applying inductive logic programming to process mining. In: Blockeel, H., Ramon, J., Shavlik, J., Tadepalli, P. (eds.) ILP 2007. LNCS (LNAI), vol. 4894, pp. 132–146. Springer, Heidelberg (2008). doi:10.1007/978-3-540-78469-2_16

16. Leemans, S.J.J., Fahland, D., van der Aalst, W.M.P.: Discovering block-structured process models from event logs containing infrequent behaviour. In: Lohmann, N., Song, M., Wohed, P. (eds.) BPM 2013. LNBIP, vol. 171, pp. 66–78. Springer, Heidelberg (2014). doi:10.1007/978-3-319-06257-0_6

17. Maggi, F.M., Mooij, A.J., van der Aalst, W.M.P.: User-guided discovery of declarative process models. In: IEEE Symposium on Computational Intelligence and Data Mining (CIDM), pp. 192–199. IEEE (2011)

18. Pesic, M., Schonenberg, H., van der Aalst, W.M.P.: Declare: full support for loosely-structured processes. In: 11th IEEE International Enterprise Distributed Object Computing Conference, EDOC, p. 287. IEEE (2007)

19. Ramezani, E., Fahland, D., van der Aalst, W.M.P.: Where did i misbehave? diagnostic information in compliance checking. In: Barros, A., Gal, A., Kindler, E. (eds.) BPM 2012. LNCS, vol. 7481, pp. 262–278. Springer, Heidelberg (2012). doi:10.1007/978-3-642-32885-5_21

20. Rembert, A.J., Omokpo, A., Mazzoleni, P., Goodwin, R.T.: Process discovery using prior knowledge. In: Basu, S., Pautasso, C., Zhang, L., Fu, X. (eds.) ICSOC 2013. LNCS, vol. 8274, pp. 328–342. Springer, Heidelberg (2013). doi:10.1007/978-3-642-45005-1_23

21. Runte, W., El Kharbili, M.: Constraint checking for business process management. In: GI Jahrestagung, pp. 4093–4103 (2009)

22. van der Aalst, W.M.P.: Process Mining - Data Science in Action, 2nd edn. Springer, Heidelberg (2016)

23. Werf, J.M.E.M., Dongen, B.F., Hurkens, C.A.J., Serebrenik, A.: Process discovery using integer linear programming. In: Hee, K.M., Valk, R. (eds.) PETRI NETS 2008. LNCS, vol. 5062, pp. 368–387. Springer, Heidelberg (2008). doi:10.1007/978-3-540-68746-7_24

Aligning Process Model Terminology
with Hypernym Relations

Stefan Bunk, Fabian Pittke, and Jan Mendling[(✉)]

WU Vienna, Welthandelsplatz 1, 1020 Vienna, Austria
bunk@ai.wu.ac.at, {fabian.pittke,jan.mendling}@wu.ac.at

Abstract. Business process models are intensively used in organizations with various persons being involved in their creation. One of the challenges is the usage of a consistent terminology to label the activities of these process models. To support this task, prior research has proposed quality metrics to support the usage of consistent terms, mainly based on linguistic relations such as synonymy or homonymy. In this paper, we propose a new approach that utilizes hypernym hierarchies. We use these hierarchies to define a measure of abstractness which helps users to align the level of detail within one process model. Moreover, we define two techniques to detect specific terminology defects, namely process hierarchy defects and object hierarchy defects, and give recommendations to align them with hypernym hierarchies. We evaluate our approach on three process model collections from practice.

Keywords: Inconsistency detection · Model quality · Process models

1 Introduction

Documenting business operations with process models has become a common practice of many organizations resulting in process collections of a considerable size. An important aspect of such collections is to keep them understandable for all stakeholders and free of contradictions and inconsistencies. This is difficult due to the size of these collections [14, 26]. Quality management therefore has to rely as much as possible on automatic, computer-assisted analysis.

While the structural information of business process models has been intensively studied [18, 28], recent research has focused on the quality of the natural language in these models [12, 17]. Recent techniques support the analysis of grammatical structures [13, 15] or the detection of semantic ambiguities [6, 25]. While the latter set of techniques use synonymy and homonymy, the potential of using other semantic relations is not well understood [16]. Specifically, hierarchy information, as given by so called hyponyms and hypernyms, are not considered so far and impose a notable gap of linguistic analysis techniques.

© IFIP International Federation for Information Processing 2017
Published by Springer International Publishing AG 2017. All Rights Reserved
P. Ceravolo and S. Rinderle-Ma (Eds.): SIMPDA 2015, LNBIP 244, pp. 105–123, 2017.
DOI: 10.1007/978-3-319-53435-0_5

In this paper, we propose a novel approach that uses hypernym relations to analyze process models and find inconsistencies. Hypernyms, or the opposite hyponyms, are words that have a broader meaning than other words [24]. An example of a hypernym-hyponym relationship is the verb pair *create* – *manufacture*. While *create* is an abstract word with many possibilities of making an object, *manufacture* captures the fact that the object is created with the hands. Our proposed technique will make use of word hierarchies to determine their level of abstraction in a given business process. Based on this, we identify two potential defects in a process model: First, we detect process hierarchy defects, which occur when a verb and one of its hypernyms is used in the same process model. Second, we identify object hierarchy defects that affect business objects which occur with verbs from the same hypernym hierarchy. Afterwards, we discuss means to resolve the detected defects in process models. In order to demonstrate the capabilities of our approach, we use three process model collections from practice and evaluate the proposed technique with them.

The rest of the paper is organized as follows. In Sect. 2, we illustrate the problem at hand and discuss previous approaches to tackle this issue. Then, Sect. 3 defines the necessary concepts of this work, before Sect. 4 introduces the details of our approach. In Sect. 5 we evaluate the metric on three real-world datasets and find defects in the data, showing the applicability of our approach. Finally, Sect. 6 concludes the paper.

2   Background

In this section, we discuss the background of our research. First, we illustrate the terminology problem based on two process models. Then, we will discuss existing research approaches for refactoring text labels in business process models.

2.1   Problem Statement

The problem of inconsistent terminology is best explained with an example. Figure 1 depicts two simple process models: *Bug Report* (Process A) and *New Feature* (Process B). Process A starts with the receipt of a bug report. Then, an employee of the first-level-support tries to reproduce the reported bug. If successful, the bug is delegated to a developer trying to find its cause. Subsequently, the developer implements a bug fix and changes the documentation which terminates the process. If the bug could not be reproduced, the process also terminates. Process B depicts the necessary steps to include a new feature in a software. After a suitable use case has been identified, the software is changed and its documentation updated which also resembles the end of the process.

The depicted process models have several problems with regard to their terminology. First, we observe that the actions in process A do not seem to be on the same level of detail. While *reproduce*, *delegate*, and *implement* are rather specific, the actions *change* and *find* refer to tasks on a higher level of abstraction.

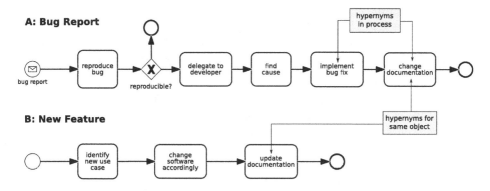

Fig. 1. Example for business processes with terminological conflicts

Thus, these actions might point to a wider range of possible tasks and leave the modeler with much space of interpretation. Depending on the required abstraction level, the modeler may wish to adjust the wording (e.g. *identify* instead of *find*), to group some activities together, or to split them into more activities.

Second, we observe a mismatch in the abstraction level that relates to the usage of verbs. In Process A, the verb *implement* is a hyponym of the verb *change* because the implementation of something always involves that a process or a product is changed. We also observe this problem between several process models of a process model collection. In the example, Process A and B both use the business object *documentation* with, however, two different actions, i.e. *change* and *update*. Similar to the previous case, *update* is a more specific expression of *change*. For the reader, it might be unclear, whether these two actions refer to the same task or not.

The problem of inconsistent terminology has been highlighted a.o. in the research of Kamsties [10] or Berry et al. [2], who identify syntax, semantic, and pragmatic ambiguities as the major source of unsound software requirements. In consequence, such unsound software requirements may lead to misconceptions within the organization with regard to the wrong selection of systems and components or the wrong implementation of system functionality [19]. Finally, as pointed out by Denger et al. [4], the correction and re-implementation will lead to expensive reworks and software deployments delays.

2.2 Related Work

Research in the field of requirements engineering has brought forth a plethora of approaches to improve the terminology of requirements documents (see Table 1). These approaches particularly focus on the issue of ambiguity detection. One option is to manually check requirements techniques by employing specific reading techniques, such as inspection-based reading [11], scenario-based reading [10],

Table 1. Overview of ambiguity management approaches

Category	Approach	Author
Ambiguity Detection in Documents	Inspection-based Reading	Kamsties et al. [11]
	Scenario-based Reading	Kamsties [10]
	Object-oriented Reading	Shull et al. [27]
	Metrics on Understandablility, Consistency, and Readability	Fantechi et al. [5]
	Ambiguity Metric	Ceccato et al. [3]
	Sentence Patterns for Ambiguity Prevention	Denger et al. [4]
	Ambiguity Detection with Regular Expressions and Keywords	Gleich et al. [8]
Ambiguity Detection in Process Models	Ambiguity and Specificity Measurement Metrics	Friedrich [6]
	Verifying Concept Relations with a Lexicon	Van der Vos [29]
	Consistency Verification with Semantic Annotations	Weber et al. [30]
	Preventing Ambiguity with a Domain Thesaurus	Becker et al. [1]
	Synonym Correction	Havel et al. [9]
	Detection and Correction of Semantic Ambiguity	Pittke et al. [25]

or object-oriented reading [27]. Besides the reading techniques, there are also automatic approaches. These make use of metrics to evaluate the requirements document based on its understandablility, consistency, and readability [5] or on its ambiguity [3]. Another class of approaches uses natural language patterns to manage ambiguity in requirements documents. Among them, Denger et al. [4] propose generic sentence patterns to describe events or reactions of a software. These patterns provide a template for a specific situation which are instantiated with the necessary concepts of the software to be developed. Gleich et al. [8] use regular expressions and keywords to detect ambiguities. For example, the expressions *many* or *few* might point to vaguely formulated requirements.

There are also techniques that specifically focus on natural language labels in process models. Friedrich [6] defines a semantic quality metric to identify activity labels that suffer from ambiguity. For that purpose, the author employs the hypernym relation of WordNet and the depth of a term in the hypernym hierarchy. The metric punishes terms if their depth is too low or too high. Van der Vos et al. [29] uses a semantic lexicon to check the quality of model elements. It ensures that words of element labels are used in a linguistically meaningful way. The technique checks if the label correctly refers to a specific object and if the combination between several objects makes sense in the context of a model. Weber et al. [30] propose a semantic annotation approach that enriches activities with preconditions and effects and propagate those for semantic consistency verification. For example, it is only possible to send a cancellation of a purchase order if it has not been confirmed yet. The approach of Becker et al. [1] enforces naming conventions in process models. Based on a domain thesaurus, the tool is capable of proposing alternative terms and preventing naming conflicts. The research prototype of Havel et al. [9] also corrects synonym terminology by selecting the dominant synonym among a set of ex-ante defined synonyms. Pittke et al. [25] automatically find synonyms and homonyms in a collection and propose a resolution with the help of word sense disambiguation and BabelNet.

The aforementioned approaches have mainly two shortcomings. First, many approaches rely on text fragments taken from a grammatically correct natural language text. However, the activity labels of process models contain only short text fragment that do not even resemble a grammatically correct sentence [15]. In addition, they follow different language patterns that can hide the action of a specific activity [17]. Therefore, the approaches for requirements documents are not directly applicable for process models. Second, most of the approaches address a particular type of ambiguity to ensure the terminological consistency, i.e. ambiguity caused by synonym and homonym usage of terms. However, the inconsistencies depicted in Fig. 1 will not be detected with them. The reason is that two words suffering from a hypernym conflict are not necessarily synonyms or homonyms such that the inconsistency is not detected at all.

Against this background, we propose an approach that explores the hypernym hierarchies between terms and uses these relations to detect inconsistencies related to the level of abstraction and to word usage.

3   Preliminaries

The detection of hypernym inconsistencies requires a basic assumption about the syntactic structure of process model activity labels. Although there is no restriction on the syntactic structure of activity labels, research has identified a set of reoccurring activity labeling styles [12, 17]. Based on the analysis of more than 1400 models of 6 collections from different industries [13], activities may be described in the verb-object, the action-noun style, the descriptive style, or an arbitrary style. Despite this variety, there are two essential components that form the nucleus of each activity label, i.e. an action and a business object. The action specifies a task that needs to be conducted in the process models, while the business object names the target of the task. Based on these insights, the our approach also assumes that each activity uses these two components.

Given a specific process model p of a process model collection P, we denote that a process model comprises a set of specific activities A_p. While activities

$a_p \in A_p$ can be formulated in different labeling styles, they contain two essential components [17], i.e. an action (a_p^A) and a business object (a_p^{BO}) on which the action is applied. As an example, consider the activity label *Reproduce bug* from Fig. 1. It contains the action *to reproduce* and the business object *bug*. It is important to note that these components can be communicated in different grammatical variations. For instance, the label *Bug reproduction* contains the same components as *Reproduce bug*, but uses a different grammatical structure. In order to be independent of grammatical structures, we use the technique of Leopold [12] to automatically extract these components. Furthermore, we assume actions to be captured as verbs and business objects as nouns.

In order to determine the hypernym/hyponym relation between words, we use BabelNet. BabelNet [22] is a large multi-lingual database of words and their senses. It combines data from WordNet [20], Wikipedia, and OmegaWiki, along with multilingual features. BabelNet organizes words in so called synsets, i.e. sets of synonymous words. Each synset describes one particular meaning of a word, which we refer to as *word sense*. These word senses are organized in an enumerative way, which means that they have been defined disjoint to each other [21]. As an example, the word senses of the verb *to reproduce* are given as follows:

- s_1: Make a copy or equivalent of
- s_2: Have offspring or produce more individuals of a given animal or plant
- s_3: Recreate a sound, image, idea, mood, atmosphere, etc.
- s_4: Repeat after memorization

Formally, we refer to these word senses as given by the following definition:

Definition 1 (Word Senses). *Let S denote the set of word senses and let W be the set of all words. Furthermore, let $POS = \{verb, noun\}$ be the set of part of speech tags. Then, the possible senses of a given word and a given part of speech tag are given by the function $Senses_{ENUM} : W \times POS \rightarrow 2^S$.*

A first important implication from this sense-based view is that hierarchies do not work on words themselves, but rather on word senses. For example, depending on the context, the verb *to destroy* may have the hypernym *to undo* or *to defeat*. Therefore, before retrieving the hypernyms from the BabelNet database, the respective word sense of the word has to be determined in the current context. This problem is known as *word sense disambiguation* (WSD). We employ the multi-lingual word sense disambiguation method by Navigli and Ponzetto [23]. WSD approaches typically require a target word, a POS tag and a context of a word for the disambiguation [21]. In our setting, we might use all the components of activity labels of a process model to perform the disambiguation task. The POS tag is determined as mentioned before: actions are verbs and business objects are nouns. The output of the sense disambiguation is a set of most likely word senses for the respective target word. WSD algorithms return a set of multiple senses if two or more senses fit the current context. We formalize the word sense disambiguation process as follows:

Definition 2 (Word Sense Disambiguation). *Let A_p be the activities of a process model and $a_p \in A_p$ a specific activity of p. Further, let $w \in \{a_p^{BO}, a_p^A\}$ be a word of the label components along with its part of speech tag pos $\in POS$, such that $S_w = Senses_{ENUM}(w, pos)$ describes the available word senses of w. Then, let $C = \{a_p^{BO}, a_p^A \mid a_p \in A_p\}$ denote the set of all words given by the label components of the process model's activities A_p. The word sense disambiguation function $WSD : W \times C \to 2^{S_w}$ selects a subset of the available senses, such that each sense describes the meaning of the word w in its context.*

We explain the definition by referring to the verb *to reproduce* from Process A in the motivating example. The senses of the verb have been listed above and together form the set $Senses_{ENUM}(reproduce, verb) = \{s_1, s_2, s_3, s_4\}$. By leveraging the context, which contains the words *bug, developer, implement* and *fix*, we are able to correctly disambiguate the third sense as the only correct one, i.e. $WSD(reproduce, C) = \{s_3\}$.

WSD enables us to explore the hypernym relation between two words, or more specifically between two word senses. In lexical databases, a hypernym relation is typically indicated if two senses are directly connected with each other. Thus, the hypernym relation can be described as follows:

Definition 3 (Hypernyms). *Given a set S of all senses, the BabelNet hypernym relationship $Hyper_{BN} \subset S \times S$ is a relation, such that $(s_1, s_2) \in Hyper_{BN}$ iff. s_1 is an immediate hypernym of s_2.*

We recursively define the function $Hypernyms : S \to 2^S$, which contains the set of all hypernyms of a sense, such that a sense $s_h \in Hypernyms(s)$ iff. $(s_h, s) \in Hyper_{BN}$ or $\exists s_m \in S : s_h \in Hypernyms(s_m) \land s_m \in Hypernyms(s)$.

Let us again consider the word *reproduce* with the disambiguated sense s_3 from Process A. Following the hypernym hierarchy step by step we find $Hypernyms(s_3) = \{s_{3,h1}, s_{3,h2}, s_{3,h3}\}$:

- $s_{3,h1}$: Re-create, form anew in the imagination, recollect and re-form in the mind
- $s_{3,h2}$: Create by mental act, create mentally and abstractly rather than with one's hands
- $s_{3,h3}$: Make or cause to be or to become

As another example from Process A, the $Hypernyms(s)$ set of the verb *change*, disambiguated as *cause to change, make different, cause a transformation*, is the empty set, because no hypernyms of that verb sense exist. In such a case, we have encountered a word that describes a general concept covering all the other concepts. Opposite to that, we might find words which describe very specific concepts that cannot be refined further. Examples of such words would be the words *to implement* or *to subscribe*.

Following this argument, we define root hypernyms and leaf hyponyms.

Definition 4 (Root and Leaf Hypernyms). *Given a sense $s \in S$ we define:*

$$RootHypernyms(s) = \{s_r \in Hypernyms(s) | \nexists s' : (s', s_r) \in Hyper_{BN}\}$$
$$LeafHypernyms(s) = \{s_l \in Hyponyms(s) | \nexists s' : (s_l, s') \in Hyper_{BN}\}$$

It is important to note that a hypernym hierarchy does not necessarily consist of a direct line to a root hypernym. A sense may have two hypernyms, whose hypernym hierarchies join again later, or it may have two different root hypernyms. Thus, there might be several senses resulting in a set of hypernym or hyponym senses. For example the verb *to execute* in the sense of *to put sth. in effect* has two root hypernyms in the BabelNet hierarchy, i.e. the senses *cause a transformation* and *make or cause to be or to become*. Leaf hypernyms of *to execute* are for example *applying oneself diligently, bringing something to perfection*, or *achieving a greater degree of success than expected*.

Finally, we need to retrieve the word corresponding to the identified hypernym senses from the previous definition. Taking the definition of word senses (Definition 2) into account, we observe that each word is linked to a set of senses. In other words, each sense might be expressed with a specific set of words. This relation is defined as follows:

Definition 5. (Words of a Given Word Sense). *Let $s \in S$ be a specific word sense from the set of all senses and $POS = \{verb, noun\}$ be the set of part of speech tags. Then, the possible words W of a given word sense and a given part of speech tag is given by the function words $: S \times POS \rightarrow 2^W$.*

Applying this definition to the previous examples of root hypernyms, we retrieve the word *to change* from the sense *cause a transformation* and *to create* from the sense *make or cause to be or to become*. We receive the words *to ply*, *to consummate*, or *to overachieve* for the exemplary senses of leaf hypernyms.

4   Conceptual Approach

In this section, we use the hypernym relationship to measure the abstractness of a process model and to identify two possible defects in business processes, i.e. *process hierarchy defects* and *object hierarchy defects*. Moreover, we explain how the detected deficiencies may be resolved in a subsequent step.

4.1   Measuring Verb Abstractness

In order to explain our measure for verb abstractness, we use the example hypernym hierarchy in Fig. 2. Intuitively, the verb *to get* is more abstract than the verb *to repurchase*, while both can describe the act of getting into possession of something. This is also represented in the hypernym hierarchy of the verb *to get* which is at a higher hierarchy level than the verb *to repurchase*. Thus, our definition of verb abstractness has to explicitly consider the position of the verb in the hierarchy, which is given by its depth and height. In our operationalization, we require both depth and height, because one concept alone does not suffice to assess the position of the node in the hierarchy. This follows the intuition that a leaf hyponym at depth 5 should be considered as more concrete than a hyponym, which is also at depth 5 but has 5 sub-levels of hyponyms below.

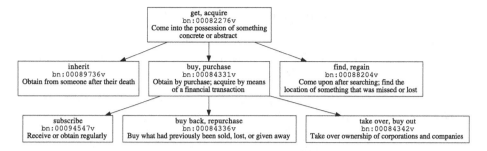

Fig. 2. Example extract from the BabelNet hypernym hierarchy. The first row shows the verbs in the synset, the second line the synset id, and the last line shows the gloss of that synset.

Definition 6 (Sense Abstractness). *For a sense* $s \in S$, *we measure the abstractness.*

$$abstractness : S \rightarrow [0, 1]$$

$$abstractness : s \mapsto \frac{height(s)}{depth(s) + height(s)},$$

such that

- *depth(v) is the length of the longest path to a root hypernym starting at s and only following hypernym relations.*
- *height(s) is the length of the longest path to a leaf hypernym starting at s and only following hyponym relations.*
- *The abstractness value is 1 for root hypernyms and 0 for leaf hypernyms.*

If we again look at the verbs from above, we find $abstractness(s_{get}) = 1.0$ and $abstractness(s_{repurchase}) = 0.0$. The word *to buy*, which lies between *get* and *repurchase* in this hypernym hierarchy, is assigned $abstractness(s_{buy}) = 0.5$.

The abstractness value of a single sense is then extended to an abstractness measure for processes by first averaging the sense abstractness over all senses of the word resulting in a measure for word abstractness. Then, we average again over all words in a process model, which is shown in the following definition:

Definition 7 (Word and Process Abstractness).

$$word\text{-}abstractness : w \mapsto \frac{1}{|WSD(w)|} * \sum_{s \in WSD(w)} abstractness(w)$$

$$process\text{-}abstractness : P \mapsto \frac{1}{|P|} * \sum_{w \in P} word\text{-}abstractness(w)$$

In the processes of Fig. 1, the *process-abstractness* values are quite similar: $process\text{-}abstractness(P_A) = 0.63$ and $process\text{-}abstractness(P_B) = 0.60$.

4.2   Process Hierarchy and Object Hierarchy Defects

We have already seen in Fig. 1 that process hierarchy defects occur, when a verb and its hypernym is used in the same process. We can automatically determine this type of defect with the following definition:

Definition 8 (Process Hierarchy Defect). *Let v_1 and v_2 be two verbs in the activity labels of one process model. Further, let the sets S_1 and S_2 be the senses of these verbs determined after WSD. We say there exists a process hierarchy defect between the verbs v_1 and v_2 iff:*

$$\exists s_1 \in S_1, s_2 \in S_2 : s_1 \in Hypernyms(s_2) \vee s_2 \in Hypernyms(s_1)$$

We call the tuple (v_1, s_1, v_2, s_2) a process hierarchy defect tuple, where s_1 is the hypernym and s_2 is the hyponym.

Looking at Process A, we find a process hierarchy defect for $v_1 = change$ and $v_2 = implement$. WSD yields $S_1 = \{s_{1,1} = (\text{cause to change}; \text{make different})\}$ and $S_2 = \{s_{2,1} = (\text{apply in a manner consistent with its purpose}), s_{2,2} = (\text{pursue to a conclusion})\}$. Now, the process hierarchy defect is the fact that the second sense of *implement* (pursue to a conclusion) is a hyponym of the first sense of *to change* (cause to change), i.e. $s_{1,1} \in Hypernyms(s_{2,2})$.

Figure 1 also depicts an object hierarchy defect. This type of defect occurs, when a verb and its hypernym is used in conjunction with the same business object. Please note that this problem only occurs for verbs that occur together with business objects from different process models. We formalize this defect as follows:

Definition 9 (Object Hierarchy Defect). *Let w_{BO} be a business object, which is used with the sense s_{BO} in a business process model collection. Then, let $S_{w_{BO}}$ be the set of all verb senses, which are used in activity labels with (w_{BO}, s_{BO}) as the business object. We say there exists an object hierarchy defect for the object/sense tuple (w_{BO}, s_{BO}) iff:*

$$\exists s_1, s_2 \in S_{w_{BO}} : s_1 \in Hypernyms(s_2) \vee s_2 \in Hypernyms(s_1)$$

The example processes in Fig. 1 contain an object hierarchy defect for the object *documentation*, which is used in the same sense in both of its occurrences. The verb senses used with *documentation* are $S_{documentation} = \{(\text{cause to change}; \text{make different}), (\text{bring up to date})\}$. Here, the first sense (cause to change) is a hypernym of the second (bring up to date), i.e. $s_1 \in Hypernyms(s_2)$.

4.3   Resolving Hierarchy Defects in Process Models

The presented detection techniques identify inconsistent terminology instances in process models. These deficiencies hinder the proper understanding and sense-making of process models since the specification of verbs on different hierarchy levels point to a varying range of possible tasks and are thus open to several interpretations. These deficiencies have to be aligned in the next step by repository managers. Accordingly, repository managers can use the BabelNet system to align deficient verbs, process hierarchy defects, and object hierarchy defects.

Regarding the alignment of *deficient verbs*, the defined abstractness metrics point to particular verbs that violate the average degree of abstractness in the respective process model. This violation might either involve verbs that are too general or too specific compared to the average *process-abstractness* score. In case of too general verbs, the repository manager can employ the BabelNet system and look for hyponym alternatives. Too specific verbs may be resolved with hyponym alternatives. For example, consider the activity *Change documentation* in Process A of our motivating example. The word-abstractness score amounts to almost 1 indicating a very general verb. In order to meet the average *process-abstractness* of 0.63, the repository manager may choose the verbs *to adapt* or *to edit* as suitable alternatives from the BabelNet hypernym tree.

Regarding the alignment of *process hierarchy defects*, the detection technique identifies verb pairs within a process model that are in a hypernym relationship with each other. Typically, this defect implies that a given task already involves the doing of another. We already mentioned the verbs *to change* and *to implement* suffering from this type of defect, because the process of changing something always involves an implementation. Accordingly, the repository manager has two alternatives. On the one hand, he could merge the two activities into one since the implementation of the bug fix also involves the change of the software documentation. On the other hand, he might keep both activities and align the action of the general activity *change documentation* by replacing it with more specific alternatives, such as *to update* or *to edit*.

Regarding the alignment of *object hierarchy defects*, the detection technique retrieves pairs of process models, in which one business objects is used with different verbs being in a hypernym relation with each other. Typically, this defect causes confusion whether or not the same task has to be applied to the respective business object. Looking at our example processes, our technique identifies the activities *update documentation* from process A and *change documentation* from process B, which leave us unclear whether changing the documentation also involves the same task as updating it. Accordingly, the repository manager might either align the two actions by choosing only one action, e.g. *to update*, or by replacing the general action *to change* with one of the aforementioned alternatives.

By applying these alignment techniques, the specificity and the terminological quality of the process model and the repository will be increased and the understandability can be improved.

5   Evaluation

This section presents the results of our evaluation. We first describe the test data and then the results of measuring abstractness and detecting hierarchy defects.

Table 2. Characteristics of the test collections

	AI	SAP	TelCo
No. of Processes	949	575	774
No. of Activities	4,193	1,545	4,972
Avg. No. of Verbs per Process	4.42	2.69	6.42
No. of Unique Verb Senses	513	354	413
No. of Unique Object Senses	1,979	556	1,641
Modeling Language	BPMN	EPC	EPC
Domain	Training	Independent	Telecommunication
Terminology Quality	Low	High	Medium

5.1 Evaluation Setup

In order to achieve a high external validity, we employ three different model collections from practice. We selected collections differing with regard to standardization, the expected degree of terminological quality, and the domain. Table 2 summarizes their main characteristics. These collections include:

- **AI collection:** The models of this collection originate from *BPM Academic Initiative*[1]. The collection was built by students and lecturers in a series of projects and lectures. From the available models, we selected those with proper English labels. The resulting subset includes 949 process models with in total 4,193 activity labels. We expect this collection to have the lowest terminology quality among our three datasets, because no common guidelines or standards exists.
- **SAP:** The *SAP Reference Model* contains 575 Event-Driven process chains in 29 different functional branches [7]. Examples are procurement, sales, and financial accounting. The model collection includes 1,545 activity labels. Since the SAP Reference Model was designed as an industry recommendation with a standardized terminology, we expect a small number of terminological defects.
- **TelCo:** The *TelCo* collection contains the processes of an international telecommunication company. It comprises 774 process models with in total 4,972 activities. We assume the TelCo collection to contain more heterogeneous terminology as it is not based on a standardized glossary. In terms of number of defects, we would expect this to be between the previous two model collections.

5.2 Evaluation of Abstractness

In the first step of the evaluation, we focus on the distribution of the *abstractness* value. In Fig. 3 we show its distribution for all collections. On average, the distribution of all our test collections shows a mean of 0.52 and standard deviation

---

[1] See http://bpmai.org/BPMAcademicInitiative/.

of 0.31. These values indicate that most of the word senses of all verbs in a single process models are in the middle of the hypernym hierarchy making use of narrow or broad verbs in a balanced way. However, we also observe process models making extensive use of verbs with either a very narrow or a very broad sense It is also interesting to note that each collection appears to be equally affected by process models with either very narrow or very broad verbs. Since, these cases more unlikely and hard to spot, the affected models require further investigation and alignment based on the overall degree of abstraction within a process model. Interestingly, this phenomenon appears in all of the three collections alike and thus being independent on the assumed labeling quality of the collections. We therefore conclude that verb abstractness has not been considered in industry process models yet and emphasizes the necessity of further investigation.

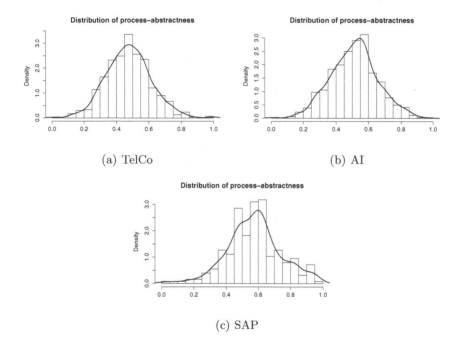

(a) TelCo (b) AI

(c) SAP

Fig. 3. Distribution of *process-abstractness* value for the test collections.

Additionally, we now look at the processes with boundary values of *process-abstractness*. Table 3 shows the verbs used in these processes along with their respective abstractness value. Following intuition, the abstract processes are dominated by words such as *to create*, *to determine*, and *to analyze*, while *to capture* and *to negotiate* appear in the concrete processes. Just by looking at the abstractness values and the verbs, it is possible to determine the level of the process in a process hierarchy.

Table 3. Abstract and concrete processes in the test collections

	Process-abstractness	Verbs in the process
TelCo	0.00	Answer, launch, test
	0.00	Capture, shop, test
	0.11	Capture, check, negotiate
	0.15	Appoint, report, review, update, validate
	0.17	Capture, check, send, store
	0.87	Change, close, create, handle, register, take
	0.87	Analyze, change, create, inform, request
	0.88	Analyze, create, initiate
	0.89	Change, determine, provide
	1.00	Create, determine, perform
SAP	0.02	Close, index, revalue
	0.06	Assign, process, requisition
	0.06	Assign, process, requisition
	0.06	Assign, process, requisition
	0.06	Post, park, receipt
	0.82	Create, determine, plan
	0.83	Analyze, direct, transfer
	0.89	Change, determine, value
	0.94	Adjust, create, determine
	1.00	Change, create, maintain
AI	0.04	Cheese, fruit, milk
	0.09	Check, return, tick
	0.11	Blaze, check, prepare
	0.11	Design, sap, test
	0.11	Click, flight, open, validate
	0.84	Access, change, choose, create, remove
	0.84	Complete, confirm, create, evaluate
	0.86	Determine, enter, search
	0.89	Close, make, solve
	1.00	Apply, close, start

5.3 Evaluation of Defects

Regarding the evaluation of hierarchy defects, we first have a look at the quantitative extent within our test collections. Table 4 lists the number of process and object hierarchy defects for each collection. Surprisingly, the TelCo collection has the highest number of process hierarchy defects per model (226 affected models with 0.85 defects on average). This might be explained by the high number

Table 4. Defects per process collection, normalized by number of processes, and number of processes with defects.

	SAP	AI	TelCo
Avg. No. of Process Hierarchy Defects	0.397	0.356	0.848
No. of Affected Process Models	72 (12.52%)	185 (19.49%)	226 (29.2%)
Avg. No. of Object Hierarchy Defects	0.171	1.620	1.303
No. of Affected Process Models	31 (5.39%)	218 (22.97%)	211 (27.26%)

Table 5. Top five detected process hierarchy defect tuples in the test collections

Count		Hypernym word and sense	Hyponym word and sense
TelCo	28	Handle: Be in charge of, act on	Manage: Watch and direct
	20	Create: Make or cause to be or to become	Initiate: Bring into being
	17	Confirm: Establish or strengthen as with new evidence or facts	Check: Make certain of something
	16	Create: Make or cause to be or to become	Plan: Make a design of; plan out in systematic form
	15	Change: Cause to change; make different	Implement: Pursue to a conclusion or bring to a successful issue
SAP	9	Create: Make or cause to be or to become	Plan: Make or work out a plan
	8	Transmit: Send from one person or place to another	Process: Deliver a warrant or summons to someone
	7	Transfer: Send from one person or place to another	Process: Deliver a warrant or summons to someone
	6	Permit: Consent to, give permission	Confirm: Support a person for a position
	5	Create: Make or cause to be or to become	Decide: Influence or determine
AI	26	Produce: Create or manufacture a man-made product	Generate: Give or supply
	21	Confirm: Establish or strengthen as with new evidence or facts	Check: Make certain of something
	19	Create: Make or cause to be or to become	Generate: Give or supply
	11	Inform: Impart knowledge of some fact, state or affairs	Prepare: Create by training and teaching
	9	Get: Come into the possession of something concrete or abstract	Receive: Come into possession of something

of verbs per process (see Table 2): the probability of process hierarchy defects rises with more verbs in a process. For the object hierarchy defects, the initially assumed terminological quality matches with the results. The SAP collection clearly has the least number of object defects per business process, followed by TelCo and AI both lying closely together.

After presenting the quantitative extent of defects in the test collections, we provide qualitative examples of word pairs that frequently cause process

Table 6. Top five detected object hierarchy defect tuples for each collection

	Rank	Word	Count	Examples
TelCo	1	Order	339	Complete – change, check – confirm
	2	Customer	100	Identify – refer
	3	Appointment	43	Plan – make
	4	Management	32	Contract – change
	5	Case	24	Generate – create
SAP	1	Order	8	Execute – complete
	2	Invoice	6	Receipt – verify
	3	Notification	5	Print – create
	4	Budget	2	Release – transfer
	5	Project	2	Schedule – plan
AI	1	Pricing options	298	Generate – create
	2	Order	138	Ship – place, process – change
	3	Claim	110	Review – evaluate
	4	Goods	100	Release – move, ship – move
	5	Application	66	Review – evaluate

hierarchy and object hierarchy defects. For that purpose, we provide tabular overviews of the five most frequent defects of each type. Table 5 gives an overview of the most frequent *process hierarchy defects* in our test collections. We, for example, identified bad combinations of the verbs *to handle* and *to manage*, *to produce* and *to generate*, or *to check* and *to confirm*. In these cases, the hypernym is conflicting with one hyponym, which causes the hierarchy defect. However, we also find hierarchy defects that involve several conflicting hyponyms. When looking at all repositories, the verb *to create* is conflicting with four different hyponyms, i.e. *to initiate, to plan, to generate, and to decide*. Even worse, the defects caused by the verb *to create* involve completely different word senses. Consider for example the TelCo collection. The verb *to create* is conflicting with the hyponyms *to initiate* and *to plan*, which emphasizes the necessity to carefully use the terminology of the model collection.

We continue with a deeper discussion of *object hierarchy defects* as depicted in Table 6. The results show that our technique is capable to uncover hypernym conflicts for business objects that are frequently used in the respective process model collection. In most of the cases, the technique sees hypernym conflicts for the processing or completion of orders, the identification and targeting of relevant customers, and the creation of pricing options. These examples illustrate that a clear distinction between the tasks applied to one object is necessary since, for example, the creation of pricing options may be done in an automatic way (*to generate*) or by a human clerk (*to create*).

Moreover, it is interesting to note that the business object *order* is involved in a notable number of object hierarchy defects throughout all test collections. For example, it conflicts with the verbs *to execute* and *to complete* in SAP or with the verbs *to check* and *to confirm* in TelCo. Both cases highlight the necessity for carefully choosing actions since the object *order* appears to play a central role in several processes. However, it is worth pointing out that the defects of the business object order in the SAP collection are far less prominent. This resembles the purpose of the SAP collection to be a an industry recommendation with a standardized terminology.

6 Conclusion

In this paper, we addressed the problem of inconsistent terminology in activity labels of business process models. We approached the problem from the perspective of hypernym/hyponym relations between verb senses. We argued that there is more freedom in choosing the verbs of a business process model, which motivates our focus on verb senses. By using the lexical database BabelNet, we operationalized the abstractness of a verb sense and used the height and the depth of the node in the hypernym hierarchy to measure the sense abstractness. Generalizing this abstractness to words and processes, we can detect abstract and concrete processes, and compare them with their level of detail in a process hierarchy if available. We use the abstractness measure to uncover two types of defects in a model collection, dealing with the usage of hypernyms. Process hierarchy defects occur if one business process contains a verb sense and one of its hypernyms, indicating different abstraction levels. Object hierarchy defects occur if one business object is used with words from the same hypernym hierarchy. The evaluation showed that these defects exist and deserve the attention of modelers.

Our technique can be used throughout the entire life cycle of a model collection. Using it during development already prevents the defects we address here. It may also increase the awareness for word selection in general and thereby increase the quality not only for verb hypernym defects. As we presented in the evaluation section, our technique can also be applied to complete collections to find ambiguities after process model creation.

In future work, we first aim to examine the effects of our method in an end user study. In such a study, we want to focus on the impact of our techniques on the understandability and maintainability of process models. Moreover, we want to gather insights and requirements from a realistic scenario that helps us to transfer our technique into existing modeling tools. Second, we also want to address the resolution of process and object hierarchy defects. The idea is to develop a recommendation-based approach that helps the modeler to correct the spotted defects in an efficient way. For that purposes, this work provides a basis for further investigation in this important research area.

References

1. Becker, J., Delfmann, P., Herwig, S., Lis, L., Stein, A.: Towards increased comparability of conceptual models - enforcing naming conventions through domain thesauri and linguistic grammars. ECIS **2009**, 2231–2242 (2009)
2. Berry, D.M., Kamsties, E., Krieger, M.M.: From contract drafting to software specification: Linguistic sources of ambiguity. Technical report, School of Computer Science, University of Waterloo, Waterloo, ON, Canada (2003)
3. Ceccato, M., Kiyavitskaya, N., Zeni, N., Mich, L., Berry, D.M.: Ambiguity identification and measurement in natural language texts (2004)
4. Denger, C., Berry, D.M., Kamsties, E.: Higher quality requirements specifications through natural language patterns. In: Conference on Software: Science, Technology and Engineering, pp. 80–90. IEEE (2003)
5. Fantechi, A., Gnesi, S., Lami, G., Maccari, A.: Applications of linguistic techniques for use case analysis. Requirements Eng. **8**(3), 161–170 (2003)
6. Friedrich, F.: Measuring semantic label quality using wordnet. In: 8. Workshop der Gesellschaft für Informatik eV (GI) und Treffen ihres Arbeitskreises Geschäftsprozessmanagement mit Ereignisgesteuerten Prozessketten (WI-EPK)", Nüttgens, M. et al. (ed.) Berlin, pp. 7–21. Citeseer (2009)
7. Gerhard, K., Teufel, T.: SAP R/3 Process Oriented Implementation: Iterative Process Prototyping. Addison Wesley, Boston (1998)
8. Gleich, B., Creighton, O., Kof, L.: Ambiguity detection: towards a tool explaining ambiguity sources. In: Wieringa, R., Persson, A. (eds.) REFSQ 2010. LNCS, vol. 6182, pp. 218–232. Springer, Heidelberg (2010). doi:10.1007/978-3-642-14192-8_20
9. Havel, J., Steinhorst, M., Dietrich, H., Delfmann, P.: Supporting terminological standardization in conceptual models - a plugin for a meta-modelling tool. In: 22nd European Conference on Information Systems (ECIS 2014) (2014)
10. Kamsties, E.: Understanding ambiguity in requirements engineering. In: Aurum, A., Wohlin, C. (eds.) Engineering and Managing Software Requirements, pp. 245–266. Springer, Heidelberg (2005)
11. Kamsties, E., Berry, D.M., Paech, B., Kamsties, E., Berry, D., Paech, B.: Detecting ambiguities in requirements documents using inspections. In: Proceedings of the First Workshop on Inspection in Software Engineering (WISE 2001), pp. 68–80 (2001)
12. Leopold, H. (ed.): Natural Language in Business Process Models. LNBIP, vol. 168. Springer, Heidelberg (2013)
13. Leopold, H., Eid-Sabbagh, R., Mendling, J., Azevedo, L.G., Baião, F.A.: Detection of naming convention violations in process models for different languages. Decisi. Support Syst. **56**, 310–325 (2013). http://dx.doi.org/10.1016/j.dss.2013.06.014
14. Leopold, H., Mendling, J., Günther, O.: Learning from quality issues of BPMN models from industry. IEEE Softw. **33**(4), 26–33 (2016). http://dx.doi.org/10.1109/MS.2015.81
15. Leopold, H., Smirnov, S., Mendling, J.: On the refactoring of activity labels in business process models. Inf. Syst. **37**(5), 443–459 (2012)
16. Mendling, J., Leopold, H., Pittke, F.: 25 challenges of semantic process modeling. Int. J. Inf. Syst. Softw. Eng. Big Co. (IJISEBC) **1**(1), 78–94 (2015)
17. Mendling, J., Reijers, H.A., Recker, J.: Activity labeling in process modeling: empirical insights and recommendations. Inf. Syst. **35**(4), 467–482 (2010)
18. Mendling, J., Verbeek, H., van Dongen, B.F., van der Aalst, W.M.P., Neumann, G.: Detection and prediction of errors in EPCs of the SAP reference model. Data Knowl. Eng. **64**(1), 312–329 (2008)

19. Mili, H., Tremblay, G., Jaoude, G.B., Lefebvre, É., Elabed, L., Boussaidi, G.E.: Business process modeling languages: sorting through the alphabet soup. ACM Comput. Surv. (CSUR) **43**(1), 4 (2010)

20. Miller, G.A.: WordNet: a lexical database for english. Commun. ACM **38**(11), 39–41 (1995)

21. Navigli, R.: Word sense disambiguation: a survey. ACM Comput. Surv. (CSUR) **41**(2), 10 (2009)

22. Navigli, R., Ponzetto, S.P.: BabelNet: the automatic construction, evaluation and application of a wide-coverage multilingual semantic network. Artif. Intell. **193**, 217–250 (2012)

23. Navigli, R., Ponzetto, S.P.: Joining forces pays off: multilingual joint word sense disambiguation. In: EMNLP-CoNLL 2012, pp. 1399–1410. Association for Computational Linguistics (2012)

24. Oxford Dictionaries: "hypernym". Oxford Dictionaries. http://www.oxforddictionaries.com/de/definition/englisch_usa/hypernym

25. Pittke, F., Leopold, H., Mendling, J.: Automatic detection and resolution of lexical ambiguity in process models. Trans. Softw. Eng. **41**(6), 526–544 (2015)

26. Rosemann, M.: Potential pitfalls of process modeling: part A. Bus. Process Manage. J. **12**(2), 249–254 (2006)

27. Shull, F., Travassos, G.H., Carver, J., Basili, V.R.: Evolving a set of techniques for OO inspections (1999)

28. Aalst, W.M.P.: Workflow verification: finding control-flow errors using petri-net-based techniques. In: Aalst, W., Desel, J., Oberweis, A. (eds.) Business Process Management. LNCS, vol. 1806, pp. 161–183. Springer, Heidelberg (2000). doi:10.1007/3-540-45594-9_11

29. van der Vos, B., Gulla, J.A., van de Riet, R.: Verification of conceptual models based on linguistic knowledge. Data Knowl. Eng. **21**(2), 147–163 (1997)

30. Weber, I., Hoffmann, J., Mendling, J.: Beyond soundness: on the verification of semantic business process models. Distrib. Parallel Databases **27**(3), 271–343 (2010)

Time Series Petri Net Models
Enrichment and Prediction

Andreas Solti[1]($\boxtimes$), Laura Vana[2], and Jan Mendling[1]

[1] Institute of Information Business, Vienna, Austria
{andreas.rogge-solti,jan.mendling}@wu.ac.at
[2] Institute for Statistics and Mathematics,
Vienna University of Economics and Business, Vienna, Austria
laura.vana@wu.ac.at

Abstract. Operational support as an area of process mining aims to predict the performance of individual cases and the overall business process. Although seasonal effects, delays and performance trends are well-known to exist for business processes, there is up until now no prediction model available that explicitly captures seasonality. In this paper, we introduce time series Petri net models. These models integrate the control flow perspective of Petri nets with time series prediction. Our evaluation on the basis of our prototypical implementation demonstrates the merits of this model in terms of better accuracy in the presence of time series effects.

Keywords: Predictive analytics · Business intelligence · Time series · Petri nets

1 Introduction

Companies need to analyze their business processes to manage their operations and to tailor effective process-aware information systems. The amount and detail of available data on business processes has substantially increased with a more intensive usage of information systems in various domains of business and private life. Process mining techniques make use of such process data in facilitating automatic discovery, conformance analysis and operational support based on log data of actual process executions [2].

While discovery and conformance have been intensively studied recently, there exists a gap of work that approaches operational support from a time prediction perspective. The few examples in this area include a performance prediction model that captures levels of load as a context factor [9], queueing networks to model business processes with waiting lines [25], or time prediction based on transition systems and log data [4]. On the other hand, it is well established that business process performance is often influenced by periodic effects, trends and delays that can range from intra-day variance of task performance of

© IFIP International Federation for Information Processing 2017
Published by Springer International Publishing AG 2017. All Rights Reserved
P. Ceravolo and S. Rinderle-Ma (Eds.): SIMPDA 2015, LNBIP 244, pp. 124–141, 2017.
DOI: 10.1007/978-3-319-53435-0_6

a process participant to storm season in Australia multiplying lodged insurance claims [3]. Up until now, there is no model that allows us to integrate such effects with the control flow of a process.

Against this background, we introduce a formal model that combines Petri nets with the analytical power of time series analysis. In this way, we are able to directly represent periodic effects, trends and delays together with the control flow specification. Our model can be described as a specific kind of a stochastic Petri net, in which the distribution and weight of each transition is replaced with time series models. The formalism is flexible, in that it can use very simple statistical models, for example the average of durations, or also more complex time series models with seasonality and trend components. We extensively evaluate this model in synthetic settings and with a case study from real-life. This current paper supersedes our earlier work [23].

The remainder of this paper is structured as follows. Section 2 presents an introductory example and summarizes prior research on predictive models for operational support in business processes. The formal model with its semantics and the methods to enrich it is presented in Sect. 3. Then, in Sect. 4, we present the evaluation setting and results. Finally, we conclude in Sect. 5 and outline challenges for future research.

2 Background

This section discusses the background of our research. Section 2.1 presents an example to illustrate the problem. Section 2.2 summarizes prior research that combines Petri nets with time. Finally, Sect. 2.3 classifies approaches to time series analysis.

2.1 Illustrative Example

Business processes are subject to seasonality [4] and effects of delaying [25]. Such effects can be modeled as time series. Figure 1 shows a respective time series for the number of airline passengers per month in thousands. This dataset is from the textbook by Box et al. [6]. Note that there is an upwards trend denoting an overall increase. Also seasonal effects are visible in this data, which hints at more or less busy periods in the calendar year.

Let us consider a travel agency that operates a call center to handle airline passenger bookings in this seasonal setting. A corresponding Petri net model is depicted in Fig. 2. It shows a call center process from the view of the customer calling. Places are depicted as circles and transitions as boxes. We have two kinds of transitions in this model. The transitions depicted as white boxes correspond to process events (e.g., a customer call is received, the voice receiving unit is left, the service ended). These transitions signal a change in the process state and correspond to progress of the case. The gray transitions are *invisible* to the system. When a customer calls, the voice receiving unit takes the call and provides routing to the corresponding service station. Customers can hang up,

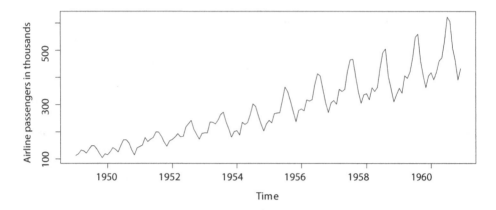

Fig. 1. Monthly airline passenger counts in thousands [6]. The data shows a clear trend and also a yearly seasonal component can be observed.

or be routed forward. Depending on whether the service station is busy, the customer needs to enter a queue first. If the customer is tired of waiting, they can hang up. They can also finish waiting in the queue to be served. Finally, the customer is connected to a service employee and when the service is finished, the process terminates.

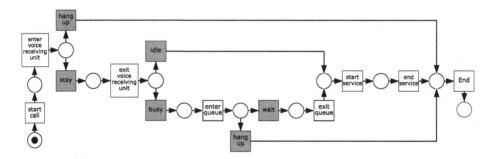

Fig. 2. Call center process as a Petri net.

Throughout this paper, we will use the Petri net formalism for modeling and prediction of business processes, more precisely the specific class of *workflow nets* [1]. We always assume the workflow net properties in this paper, and use the term Petri net instead. Petri nets are a versatile tool that allow us to capture behavioral relations, like concurrency, conflict and sequential behavior. They also allow us to capture repeated cycles in a process in a compact form. All common process modeling languages, no matter if *imperative* like BPMN, EPC, or UML Activity Diagram [15], or *declarative* such as Declare models [20] can be mapped to Petri nets. Therefore, by building on Petri nets, we effectively offer a means to predict durations for any model that has a Petri net representation.

2.2   Petri Nets and Time

Although there has been extensive research on combining Petri nets with time, there is until now no model available that directly integrates it with time series characteristics. However, various extensions to Petri nets have been proposed to capture the non-functional properties—like temporal performance—of systems.

After various approaches to incorporate fixed temporal aspect (e.g., fixed temporal durations for transitions, or interval bounded durations) researchers tried to better accommodate the fact that durations often have a stochastic nature. One of the first extensions in this direction was to enrich each transition in a Petri net with exponential firing rates and the resulting models are called *stochastic Petri nets* (SPN) [16]. These models are memory-less in their firing behavior. That is, their behavior is independent of the time spent in a certain state. This property makes SPN isomorphic to Markov chains [19]. To overcome this simplification and allow to natively model non-exponential durations, *non-Markovian* stochastic Petri nets were proposed. Latter allow for general modeling of the duration distributions of transitions [10]. All these models assume independence of durations within a process instance, and also between cases. An extension to capture dependence on the history of the current case was introduced in the notion of History dependent stochastic Petri nets [24]. Latter models allow us to capture an often encountered phenomenon in business processes with cycles: the probability to leave a cycle increases with each iteration. Note that Petri nets allow us to capture the dependence between cases and processes, by modeling all the cases and processes in one system. In this scenario, resources (e.g., human process participants, machines) can be represented as tokens that synchronize shared activities or create realistic queueing for scarce resources. Example works in this direction are from van der Alast [26] or the textbook by Zhou and Venkatesh [27]. Even queueing theory found its way into Petri nets by [5], where places in the net are representing queueing stations.

Besides Petri net based models, more abstract models building on transition systems were also proposed to predict remaining process durations [4]. These type of approaches extract a state s from the given observed process trace, and predict the average remaining durations of former cases that also passed through state s. Extensions to make these methods more accurate have been devised, for example to *cluster* cases based on the system load (i.e., the number of currently active process instances of a process) [9]. Another approach to predict the remaining time is based on feature-based regression of different characteristics of the case and works well, if the features are correlated to the remaining duration [7]. These methods work well, if process data is available to the process engine, and an extension of our approach with regression is certainly worthwhile to investigate, but out of scope for this paper.

These models, however, are unable to capture seasonality and trends in data. To our knowledge, we present in the following the first work integrating time series and Petri nets.

2.3 Time Series

Time series data arise naturally when monitoring processes over a given period of time. Time series analysis methods have the advantage of accounting for the fact that data observed over time might have an underlying internal structure. Hence, the goal of time series analysis is the understanding of this underlying structure and of the forces driving the observed data as well as forecasting of future events. More formally, a time series is defined as a sequence of observations at given times. We use the following notation to describe the past N observations: $y_1, \ldots, y_N$. Further, we are interested in the value of a time series h steps ahead in the future, that is, we predict y_{N+h}. The parameter h is called *horizon*, as it marks how far we would like to look ahead into the future to predict the parameter in question.

Observed time series data can be decomposed into several potential components: a random or shock component, a trend component (a systematic linear or non-linear tendency in the time series), and a seasonal component (patterns that repeat themselves in systematic time intervals). Other patterns in time series data might include autocorrelation (correlation with different lags of the data), also known as autoregressive AR processes, correlation with different lags of the shocks, called moving average (MA) processes, or both. Latter are called ARMA processes. One property necessary for predicting and modeling time series is stationarity, which requires a constant mean of the series. In general, non-stationary data is transformed to stationary data by differentiation.

There are different techniques for modeling and forecasting time series data [11]. The most popular ones include Exponential Smoothing and the Box-Jenkins ARIMA (AutoRegressive Integrated Moving Average) models, which except for incorporating AR and MA patterns can also account for seasonality components. Further, there are several naive approaches to forecasting, e.g., simply using the last observed value of the current time series as forecast. The large body of research dealing with forecasting for time series is concisely summarized in the review by de Gooijer and Hyndman [11].

In the following, we will denote as **Y** the universe of time series models, i.e., models that when provided with a given time series $\{y_1, \ldots, y_N\}$ can be used to generate a prediction for the given forecast horizon. Note that we will not restrict the kind of models that can be used to certain kinds of time series models. In fact, in the evaluation, we will compare different approaches, from naive predictors to the automatic selection of a fitting ARIMA model.

3 Time Series Petri Nets

In this section, we describe the underlying model that we propose for encoding seasonality and trends in Petri nets. Consequently, we can use these enhanced model for example to predict remaining time of business processes. Section 3.1 introduces the time series Petri net (TSPN) model. Section 3.2 discusses the challenges of enriching Petri nets towards TSPN models. Section 3.3 clarifies our design decisions.

3.1 Definition and Semantics

In contrast to previous approaches that use Petri nets for performance modeling, we allow the model to encode correlations to previous observed values at given stations in the process, i.e., at the transitions. In this sense, the model we propose builds on the one of Schonenberg et al. [24], into which we integrate time series concepts such as correlations to previous instances which passed through the same part of the model. This allows us to capture seasonality in durations and decisions. For example, at Christmas more customers choose the gift wrapping option in the order process than normally. The choice between conflicting transitions is captured by dynamic weights which depend on time and on the previous trends or seasonal patterns that can be observed.

Given a plain model of a Petri net model as $PN = (P, T, F, M_0)$, the TSPN is a model is defined as follows:

Definition 1 (Time Series Petri Net). *A TSPN is a six-tuple: TSPN = $(P, T, F, M_0, \mathcal{C}, \mathcal{D})$, where (P, T, F, M_0) is the basic underlying Petri net.*

- *P is a set of places.*
- *The set of transitions $T = T_I \cup T_T$ is partitioned into immediate transitions T_I and timed transitions T_T*
- *$F \subseteq (P \times T) \cup (T \times P)$ is a set of connecting arcs representing flow relations.*
- *$M_0 \in P \to \mathbb{N}_0$ is an initial marking.*
- *$\mathcal{C} : T_I \to \mathbf{Y}$ assigns to the immediate transitions their corresponding time series model that represents their firing rate (which can vary over time).*
- *$\mathcal{D} : T_T \to \mathbf{Y}$ is an assignment of time series models to timed transitions reflecting the durations of the corresponding process states.*

This definition of TSPN models is aligned with the well-established generalized stochastic Petri net (GSPN) [18] model, where immediate transitions are responsible for routing decisions, and timed transitions represent durations between states in the process. The execution semantics of the TSPN model can be chosen from the combinations of conflict resolution and firing memory policies, as it is also available for non-Markovian stochastic Petri nets [17]. Without loss of generality, we assume that conflicts between immediate transitions are resolved probabilistically with respect to their estimated firing rates as forecast by their time series, and conflicts between concurrently enabled timed transitions are resolved by a race policy, that is, the fastest transition fires first. Those transitions that lose a race can keep their progress (i.e., we use the enabling memory policy) until they get disabled. Note that this choice is not binding, and other execution semantics (e.g., the preselection method), as discussed by Marsan could be substituted [17]. The approach does not specifically depend on a specific policy.

3.2 Challenges of Enriching Time Series Petri Nets

The enrichment process is closely following the algorithm as described for generally distributed transition stochastic Petri nets [21]. In a nutshell, the event log that contains the collected execution information of a process is replayed on the corresponding Petri net model. This Petri net can be either manually provided, or discovered from the event log by process mining techniques [2]. During the replay, we obey the semantics of the TSPN model, which are flexibly selected by the user. This way, we gather for every transition the time at which it fired, and also the time at which another conflicting transition fired instead. This information can be used to estimate the firing rates of transitions which possible change over time. The result is an enriched Petri net, where for each transition we collected the *durations* from enabling to firing with the associated timestamp of firing. This information then can be used to train (possibly time-dependent) models.

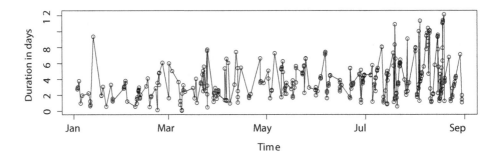

Fig. 3. Irregularly spaced time series of a logistics process. The duration until a container is picked up is depicted.

Various challenges and modeling options have to be considered for the construction of an appropriate model. We encountered the following challenges for enriching Petri net models to TSPN models.

Immediate Transitions. Without prior knowledge, transitions in Petri net models can be either immediate or timed transitions. Only by careful analysis of the durations can we decide whether a transition is immediate.

Irregularly Spaced Observations. Note that in contrast to common time series data, we consider durations of activities (or more generally durations of certain process states), where the gathered observations are irregularly spaced. See Fig. 3, which shows the durations for transport containers remaining at a harbor. In this case, the density of the collected observations clearly demonstrates that we collected more data points in August than in March.

Outliers. It can happen that the duration of an activity is a rather extreme value compared to the other values. There can be many possible reasons for outliers— e.g., coordination problems of process participants, rare cases that require more work than normally. Because outliers can negatively impact the accuracy of learned models, it can be beneficial to first remove outliers before learning the model parameters of a time series from data.

Decision Probabilities. In business processes that capture choices with regard to the following path in the process, decisions can be modeled as probabilistic choices. We need to find a way that allows us to capture temporal patterns and dependencies in the decision probabilities. Typically, immediate transitions are equipped with static weights that do not capture this aspect.

Hidden Patterns and Model Selection. It is no trivial task to identify the appropriate model that fits the observed data well and also generalizes to future data. Sometimes, it is surprising how complex models with many parameters are fitted to data, but do not generalize well to new observations. The difficulty lies in the balance between *overfitting* and *generalization* [12].

Negative Values. Time series models are usually agnostic of the sign of the data (be it positive or negative). That is, they usually have no mechanism to stay in the positive region. In our setting, we have durations and firing rates of transitions, which need always be positive.

Besides these challenges, we highlight an important technical detail, as it might be easily missed when constructing time series from such collected data: The collected data is recorded when the transitions *fire*. However, as the model should represent the duration of a transition in dependence of the time, we need to shift the observation to the point in time when the state was *entered*.

3.3   Design Decisions for the Enrichment of Time Series Petri Nets

In this section, we discuss possible solutions to these challenges, and the solutions we chose to implement for a prototypical evaluation.

Detecting Immediate Transitions. Our proposed solution to identify immediate transitions is to check whether the a specific percentile (e.g. the 95th percentile) of the collected values is below a user defined (small) threshold. This way, the method is robust to minor differences in system times in distributed settings, and also robust to a limited number of outliers in the data (e.g., 5%).

Avoiding Irregularly Spaced Observations. We apply a straight-forward technique to convert irregularly spaced time series into equidistant observations, which is *aggregation* to a coarser grained time unit. For example, one can aggregate the durations of a given activity to an hourly basis using the average of the observations in each hour as the aggregate value. By this transformation, seasonal patterns are easier to identify, as the averages of each morning at 10 am are 24 observations apart, with a weekly period of $24 \cdot 7 = 168$. Note that the best granularity of abstraction (e.g., hourly, daily, weekly) depends on the frequency of observations and will vary between processes, and perhaps also between different transitions in one model.

Removing Outliers and Missing Values. One way to deal with outliers is to remove them from the training data to which we want to fit the time series models. There exist ways to detect temporal outliers in business processes [22], which we could use to identify a certain number of the most extreme values. Further, there also exists an implementation in R, which we use to remove outliers and missing values from time series data.[1]

Using Time Series to Model Decision Probabilities. Our goal is to keep the model consistent. That is, we also want to be able to capture *decision probabilities* that vary over time, not only to allow for durations of activities to be dependent on time. Thus, in contrast to [21], we do not only count the number of times a certain decision was taken in comparison to the conflicting decisions, but capture the *count* of transition firings as the time series of the immediate transition. Let us consider a case of two conflicting transitions. By aggregating the counts on an hourly basis, we can determine the firing rate of these transitions in the next hour as the ratio between the two forecasts of the corresponding time series models. Thereby, we can effectively capture temporal patterns (e.g., seasonal components, trends) in the decision probabilities of TSPN models.

Identifying Hidden Patterns and Selecting Models. To create a plausible prediction model, we need to integrate domain knowledge of experts, who understand the business processes. There is no silver bullet solution for modeling the data, although the recent advancements in computational power and techniques enable us to automate parts of the analyses that statisticians do— see for example the automatic statistician research project[2]. Here, we keep the implementation of the model open such that any model that can be applied to time series data can be plugged into the transitions.

As a use case, we selected the `auto.arima()` function provided in the `forecast` package in R [14]. The `auto.arima()` function fits a number of different ARIMA time series models with varying parameters and selects the one that yields the best tradeoff in accuracy and model complexity. Additionally, we implemented further naive time series predictors to compare prediction accuracies.

Avoiding Negative Values. Imagine a negative trend in the durations of an activity—perhaps caused by a process participant getting more efficient in handling cases over time. If we simply extrapolate a negative trend when forecasting, we will eventually forecast negative duration values, which we need to avoid. To prevent this effect, whenever a model forecasts negative durations, we replace these negative forecast values with 0. Alternative solutions would be to use log-transforms data and predict in the log-space, and then to transform the predictions back by exponentiation. We did not use the latter approach, as it has problems with dealing with zero values. Zero values occur naturally in our setting, e.g., when a transition is not fired in a time unit of aggregation, the count value is 0. This would then entail more complex treatment.

---

[1] The `tsclean()` method in the `forecast` package in R provides automatic interpolation of missing values and removal of outliers.

[2] The Automatic Statistician project: http://www.automaticstatistician.com.

Now that we discussed possible solutions to the challenges, let us focus on the most critical challenge for integrating time series approaches with business process models: the challenge of irregularly spaced data points. Note that ignoring this issue and simply treating the inter-arrival times between cases at a processing step would destroy the option to capture seasonality. Thus, we seek another solution that maintains the temporal patterns and allows us to capture patterns as introduced in Fig. 1.

There is an important trade-off, which we need to keep in mind using the approach of aggregation. On the one hand, we gain efficiency, that is, we can reuse a forecast value for one hour for all the cases that need a forecast in that hour. On the other hand, we lose the patterns inside the unit of aggregation. Additionally, if we choose a too narrow time unit, we will end up with a lot of time units with missing values (time units, in which no single case was observed). Therefore, the aggregation level should be based on the expected granularity of the seasonal patterns. In the following, we shall evaluate the TSPN formalism with respect to its predictive performance in synthetic and real settings.

4   Evaluation

In the previous section, we presented a novel approach to capture temporal dependencies within business processes. To evaluate its usefulness in the business process domain, we first evaluate the model's predictive performance with synthetic process models. This way, we are able to identify possible conceptual advantages of the model given *clean* data that follows seasonal patterns. That is, we are interested in answering the question of how much better a time series model would be in comparison to models that do not incorporate temporal dependency structures, in a setting with clear temporal dependency structures. Section 4.1 defines the setting of our experiments. Section 4.2 describes the compared models. Section 4.3 shows the results. Section 4.4 discusses the merits of our model.

4.1   Experimental Setting

To conduct the experiment, we created TSPN models with 10 activities in sequence (we do not use complex control flow structures, because we want to isolate the prediction performance and not distort the results with synchronization effects). The activities have either a sinusoidal pattern (representing a seasonal pattern) of follow a random ARMA process. More specifically, the process parameters are randomly drawn according for the following processes.

The *sinusoidal* process uses the following equation given a time point t:

$$Y_t = \alpha + \gamma \cdot \sin(t \cdot \beta) + \epsilon_t \tag{1}$$

Here, α is the intercept or mean value, γ is the amplitude and β is the frequency. Additionally, the process has an attached normally distributed error term ϵ.

The *ARMA* process is generated by the following process:

$$Y_t = \sum_{i=1}^{n} \alpha_i y_{t-i} + \sum_{i=1}^{m} \beta_i \epsilon_{t-i} + \epsilon_t \tag{2}$$

This process consists of an autoregressive part with order n and parameters $\alpha_1, \ldots, \alpha_n$, and a moving average part with order m and parameters $\beta_1, \ldots, \beta_m$. It also includes a normally distributed error term ϵ.

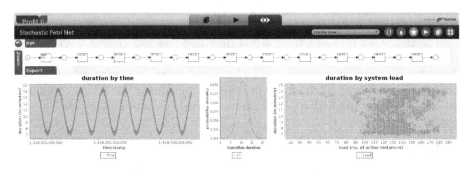

Fig. 4. Screenshot of a sequential time series Petri net with a sinusoidal pattern of durations.

We implemented the TSPN formalism in the open-source process mining software ProM[3]. The package is freely available as open-source software and the synthetic data sets are provided for testing as well. Figure 4 shows a screenshot of the plug-in visualizing an enriched sinusoidal TSPN process. The main window shows the process structure, and the lower windows project some statistical information about the durations. From right to left this is the duration plotted as a scatter plot against the system load (i.e., the number of concurrently active cases), an aggregate plot of the collected duration values as a probability density function (middle), and the activity durations of the observed cases in relation to time (left). In this lower left screen, analysts can quickly identify temporal patterns in their process data, insofar as these are present.

After the creation of the synthetic models with their corresponding processes, we sample 10, 000 process instances from each model and thereby obtain the simulated event logs. We will use these logs to test how different time series models and time agnostic models for prediction based on Petri nets [21] are able to capture the patterns in the processes.

For the evaluation, we need to ensure that our predictions are only based on available data. Notably, we cannot rely on a usual *cross-validation* approach when using time series data. Instead, we need a *rolling forecasting origin* approach. With a rolling forecasting origin, the first k observations $y_1, \ldots, y_k$ are used as training set to train the models that are then used to predict the next

---

[3] See **StochasticPetriNet** package in ProM: http://www.promtools.org/.

value with a forecast horizon h. This procedure is repeated with the next forecast using the next observations $y_{1+h}, \ldots, y_{k+h}$. In our case, we use 10% as training data and repeat this procedure for every further event in the event log.

Let us illustrate the approach with the example of predicting the duration of a single activity, which translates to predicting the firing time of the corresponding transition of the Petri net. In this case, the previous observations of that transition's duration are aggregated into a time series of the desired granularity—in our case into hourly averages. These aggregates are used as *training* data to fit the models that we want to compare.

Having fit the models, we then want to predict the duration of an activity. Therefore, we check the timestamp at the prediction time and compare it to the last observation's timestamp. The difference of the timestamps in hours is the forecast horizon h. For example, if we want to predict the duration of the service station for a customer that just called on Monday morning at 8 am, and the last observation was on Friday evening at 6 pm, then the forecast horizon h is $6 + 24 + 24 + 8 = 62$ h.

4.2 Compared Models

As mentioned in the previous section, we selected the `auto.arima()` based model, which ideally selects the best fitting representative of a family of ARIMA models. But we also implemented four naive predictors common to time series analysis [13, Chap. 2.3]. Let us denote the number of observations in the time series as N, the predicted value of a model as $\hat{y}$, and the forecast horizon as h.

The *average method* completely ignores any temporal patterns. It predicts the next observation as the average of the values observed so far and is defined as:

$$\hat{y}_{N+h|N} = \bar{y} = \frac{y_1 + \cdots + y_N}{N}. \tag{3}$$

The *naive method* uses the last observation as next predictor and ignores the horizon:

$$\hat{y}_{N+h|N} = y_N \tag{4}$$

The *seasonal naive method* (with a season-parameter m) is similar to the naive one, but it uses the observation from the last season to predict the duration:

$$\hat{y}_{N+h|N} = y_{N+h-km}, \quad \text{with } k = \left\lfloor \frac{h-1}{m} \right\rfloor + 1 \tag{5}$$

The *drift method*, allows the forecasts to linearly increase or decrease over time, where the amount of change over time (called the drift) is set as the average change of the historical data. So the forecast for time $t + h$ is given by:

$$\hat{y}_{N+h|N} = y_N + \frac{h}{N-1} \sum_{t=2}^{N} (y_t - y_{t-1}) \tag{6}$$

This is equivalent to drawing a line between the first and last observation, and extrapolating it into the future.

Further, we added two Petri net based time prediction models which do not consider seasonal or trend effects in the data. Namely, a GSPN model using exponential distributions, and a generally distributed transition stochastic Petri net (GDT_SPN) model based on a non-parametric Gaussian kernel regression for the distribution of the duration observations. These models serve as a reference for models without time series features.

4.3   Evaluation Results

We are interested in how well the models can predict the observed behavior out of sample. For each prediction iteration (i.e., when a new event is observed) we compare the predicted remaining process duration to the actual remaining duration recorded in the log. The difference between the forecast an the actual value is called prediction *error*. By aggregating the errors in certain measures, we can weight the prediction quality of the different models against each other.

Therefore, we measure the *precision* of the model as the bias, which is represented by the mean error, and look at two *accuracy* measures: The mean absolute error (MAE) and the root mean square error (RMSE). Latter is more sensitive to our model predicting values far off the observed values, while the MAE is an easily interpretable measure: It tells us that on average, our model makes an error of the size of the MAE.

Figure 5 shows the prediction results of the competing models for the sinusoidal duration pattern Fig. 5a and the ARMA-driven duration process Fig. 5b. It can be read from the plots showing the RMSEs and MAEs that the `auto.arima()` based model fits the sinusoidal pattern well in comparison to the naive methods. In the ARMA case, the `auto.arima()` method does not fully capture the pattern of the underlying process. We also see that the two naive methods (the naive method using the last observation and the drift method that adjusts the last observation with a drift) are converging. This is expected, as the prediction horizon is mostly only one step, because there are no gaps in the data.

Note that the TSPN based approaches taking into account temporal relationships outperform those comparison methods in terms of accuracy of prediction that cannot make use of the temporal patterns in the data (i.e., the GSPN, GDT_SPN, and the simple average TSPN models).

With these promising results on the synthetic data sets, we next return to our example that we introduced in Sect. 2.1. We use the Petri net depicted in Fig. 2 and a corresponding event log capturing 28 439 process instances of a call center process recorded in January 1999[4].

Figure 6 shows the prediction results for the call center case study. We conducted the same experiment as with the synthetic processes (i.e., first 10% cases as training data, then rolling forecasting for the next cases). Here, we observe that the differences of the various competing models are not substantial with

---

[4] The data of the call center process is available at http://ie.technion.ac.il/Labs/Serveng.

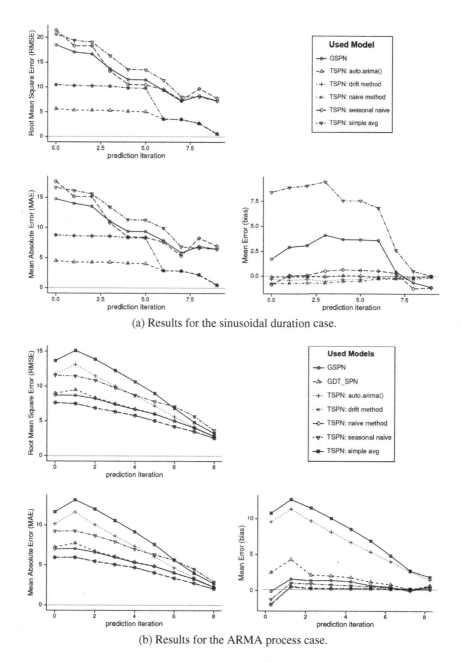

(a) Results for the sinusoidal duration case.

(b) Results for the ARMA process case.

Fig. 5. Prediction results for the sequential process with (a) sinusoidal duration patterns and (b) random ARMA processes generating transition durations. All competing models were run with the rolling forecasting origin method. The prediction is made at each new observed event based on an event log of 10,000 cases and a rolling forecasting window of 1,000 cases.

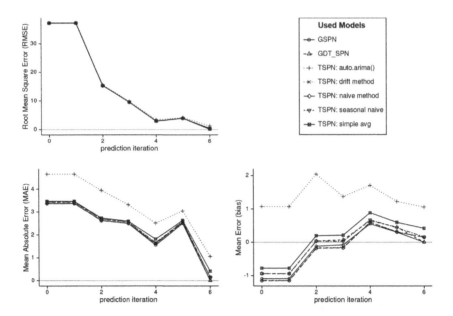

Fig. 6. Prediction results for the call center case study. All competing models were run with the rolling forecasting origin method. The prediction is made at each new observed event for the model in Fig. 2 and an event log containing 28,439 cases from January 1999.

the RMSE values of the predictors being in a similar range. This indicates, that this process does not have a temporal autocorrelation on hourly aggregate values that the compared forecast methods could exploit. However, it has to be noted that all time series predictors except for the `auto.arima()` perform equally well even without time series effects in the data. Latter creates a positive bias in the predictions and overestimates the durations, as can be seen on the graph for the right hand side in Fig. 6.

4.4 Discussion of the Results

The results presented here can be summarized as follows. In the presence of time series effects as in the synthetic logs, our TSPN models can effectively exploit these for making better predictions than standard stochastic Petri nets. In the absence of time series effects as in the real-life log case study, time series models perform equally well as the baseline. We observe that the `auto.arima()` function appears to be less robust in this case with partially creating biased estimates. Furthermore, it has to be noted that the temporal granularity might have an influence on the visibility of time series effects. For the case of the real-life data, we worked with hourly aggregates. At this stage, we cannot rule out that time series effects might be visible on a finer level. General guidelines for the choice of a specific aggregation level depend on the frequency of observations, but have to be inspected in future research.

5   Conclusion

With this paper we introduced the first model that integrates seasonal aspects and trends with the control flow structure of business process modeling. We provided the formal model with its semantics, enrichment and an open-source implementation accompanied by the synthetic test data.

There remain some open research challenges to analyze in future work. For example, the results in the selected case study imply that there is potential for improvement on how to best capture the exhibited patterns. One branch of future research is to investigate more sophisticated methods in time series that are able to capture irregularly spaced time series data [8]. Another branch is to automatically find the appropriate abstraction levels per transition. To tackle the overfitting issue, coarser grained time series models can be used and combined with the finer grained models in a weighted average fashion.

We feel that the potentials of time series methods are not yet fully unleashed with our approach. Different parameter settings influence the predictions. Optimal selection of parameters and time series models is an open challenge. To further increase the accuracy of predictions, the timetables and availabilities of resources could be captured. For that purpose, we envision an automatic integration of holidays and of working schedules into the models. Alternatively, an integration with resource-aware models (e.g., discovered queueing networks [25] would combine resource-based and seasonal effects.

Acknowledgement. This work was partially supported by the European Union's Seventh Framework Programme (FP7/2007-2013) grant 612052 (SERAMIS).

References

1. van der Alast, W.M.P.: The application of petri nets to workflow management. J. Circ. Syst. Comput. **8**(1), 21–66 (1998)
2. van der Alast, W.M.P.: Process Mining - Discovery, Conformance and Enhancement of Business Processes. Springer, Heidelberg (2011)
3. van der Alast, W.M.P., Rosemann, M., Dumas, M.: Deadline-based escalation in process-aware information systems. Decis. Support Syst. **43**(2), 492–511 (2007)
4. van der Alast, W.M.P., Schonenberg, H., Song, M.: Time prediction based on process mining. Inf. Syst. **36**(2), 450–475 (2011)
5. Bause, F.: Queueing petri nets-a formalism for the combined qualitative and quantitative analysis of systems. In: 5th International Workshop on Petri Nets and Performance Models, Proceedings, pp. 14–23. IEEE (1993)
6. Box, G.E., Jenkins, G.M., Reinsel, G.C., Ljung, G.M.: Time Series Analysis: Forecasting and Control, 4th edn. Wiley, New York (2013)

7. Dongen, B.F., Crooy, R.A., van der Aalst, W.M.P.: Cycle time prediction: when will this case finally be finished? In: Meersman, R., Tari, Z. (eds.) OTM 2008. LNCS, vol. 5331, pp. 319–336. Springer, Heidelberg (2008). doi:10.1007/978-3-540-88871-0_22

8. Engle, R.F., Russell, J.R.: Autoregressive conditional duration: a new model for irregularly spaced transaction data. Econometrica **66**(5), 1127–1162 (1998)

9. Folino, F., Guarascio, M., Pontieri, L.: Discovering context-aware models for predicting business process performances. In: Meersman, R., et al. (eds.) OTM 2012. LNCS, vol. 7565, pp. 287–304. Springer, Heidelberg (2012). doi:10.1007/978-3-642-33606-5_18

10. German, R.: Non-Markovian analysis. In: Brinksma, E., Hermanns, H., Katoen, J.-P. (eds.) EEF School 2000. LNCS, vol. 2090, pp. 156–182. Springer, Heidelberg (2001). doi:10.1007/3-540-44667-2_4

11. de Gooijer, J.G., Hyndman, R.J.: 25 years of time series forecasting. Int. J. Forecast. **22**(3), 443–473 (2006)

12. Hand, D., Mannila, H., Smyth, P.: Principles of Data Mining. MIT press, Cambridge (2001)

13. Hyndman, R.J., Athanasopoulos, G.: Forecasting: principles and practice. OTexts (2014). https://www.otexts.org/book/fpp

14. Hyndman, R.J., Khandakar,Y.: Automatic time series for forecasting: the forecast package for R. Technical report, Monash University, Department of Econometrics and Business Statistics (2007)

15. Lohmann, N., Verbeek, E., Dijkman, R.: Petri net transformations for business processes – a survey. In: Jensen, K., Aalst, W.M.P. (eds.) Transactions on Petri Nets and Other Models of Concurrency II. LNCS, vol. 5460, pp. 46–63. Springer, Heidelberg (2009). doi:10.1007/978-3-642-00899-3_3

16. Marsan, M.A.: Stochastic petri nets: an elementary introduction. In: Rozenberg, G. (ed.) APN 1988. LNCS, vol. 424, pp. 1–29. Springer, Heidelberg (1990). doi:10.1007/3-540-52494-0_23

17. Balbo, G., Bobbio, A., Chiola, G., Conte, G., Cumani, A.: The effect of execution policies on the semantics and analysis of stochastic petri nets. IEEE Trans. Softw. Eng. **15**, 832–846 (1989)

18. Marsan, M.A., Conte, G., Balbo, G.: A class of generalized stochastic petri nets for the performance evaluation of multiprocessor systems. ACM TOCS **2**(2), 93–122 (1984)

19. Molloy, M.K.: On the integration of delay and throughput measures in distributed processing models. Ph.D. thesis, University of California, Los Angeles (1981)

20. Prescher, J., Ciccio, C.D., Mendling, J.: From declarative processes to imperative models. In: Proceedings of the 4th International Symposium on Data-driven Process Discovery and Analysis (SIMPDA), pp. 162–173 (2014)

21. Rogge-Solti, A., van der Aalst, W.M.P., Weske, M.: Discovering stochastic petri nets with arbitrary delay distributions from event logs. In: Lohmann, N., Song, M., Wohed, P. (eds.) BPM 2013. LNBIP, vol. 171, pp. 15–27. Springer, Heidelberg (2014). doi:10.1007/978-3-319-06257-0_2

22. Rogge-Solti, A., Kasneci, G.: Temporal anomaly detection in business processes. In: Sadiq, S., Soffer, P., Völzer, H. (eds.) BPM 2014. LNCS, vol. 8659, pp. 234–249. Springer, Heidelberg (2014). doi:10.1007/978-3-319-10172-9_15

23. Rogge-Solti, A., Vana, L., Mendling, J.: Time series petri net models - enrichment and prediction. In: Proceedings of the 5th International Symposium on Data-driven Process Discovery and Analysis (SIMPDA), Vienna, Austria, December 9–11, pp. 109–123 (2015)

24. Schonenberg, H., Sidorova, N., van der Aalst, W.M.P., Hee, K.: History-dependent stochastic petri nets. In: Pnueli, A., Virbitskaite, I., Voronkov, A. (eds.) PSI 2009. LNCS, vol. 5947, pp. 366–379. Springer, Heidelberg (2010). doi:10.1007/978-3-642-11486-1_31

25. Senderovich, A., Weidlich, M., Gal, A., Mandelbaum, A.: Queue mining for delay prediction in multi-class service processes. Inf. Syst. **53**, 278–295 (2015)

26. van der Aalst, W.M.P.: Petri net based scheduling. Oper. Res. Spektrum **18**(4), 219–229 (1996)

27. Zhou, M., Venkatesh, K.: Modeling, Simulation, and Control of Flexible Manufacturing Systems - A Petri Net Approach. vol. 6 of Series in Intelligent Control and Intelligent Automation. World Scientific, Singapore (1999)

Visual Analytics Meets Process Mining: Challenges and Opportunities

Theresia Gschwandtner[✉]

CVAST – Centre for Visual Analytics Science and Technology,
Vienna University of Technology, Vienna, Austria
gschwandtner@cvast.ac.at
http://www.cvast.tuwien.ac.at/cvast

Abstract. Event data or traces of activities often exhibit unexpected behavior and complex relations. Thus, before and during the application of automated analysis methods, such as process mining algorithms, the analyst needs to investigate and understand the data at hand in order to decide which analysis methods might be appropriate. Visual analytics integrates the outstanding capabilities of humans in terms of visual information exploration with the enormous processing power of computers to form a powerful knowledge discovery environment. The combination of visual data exploration with process mining algorithms makes complex information structures more comprehensible and facilitates new insights. In this position paper I portray various concepts of interactive visual support for process mining, focusing on the challenges, but also the great opportunities for analyzing process data with visual analytics methods.

Keywords: Visual process mining · Visual analytics · Challenges

1 Introduction

Today we are confronted with an overload of data and information. The amount of information produced on a daily basis is just too much for anyone to process in order to answer specific questions. Examples include data from electronic health records, real time sensors, communication logs, and financial transactions. However, somewhere within this huge amounts of data, there is valuable information to solve specific tasks and answer specific questions. Yet, in order to make sense of the data, we need to find appropriate ways to process it. There are two opposed approaches to tackle this problem: (1) machine learning and data mining, and (2) visualization of the data.

The first approach, i.e. using machine learning and data mining, taps the computational power of the computer for statistical analysis of the data. As a result, a report is generated, giving - at best - the answer to the user's question. However, the second approach, i.e. visualizing the data, utilizes the power of human perception to simultaneously process large amounts of data. When visualizations are carefully designed (e.g., using a suitable combination of preattentive visual attributes to encode important information), the human perception

© IFIP International Federation for Information Processing 2017
Published by Springer International Publishing AG 2017. All Rights Reserved
P. Ceravolo and S. Rinderle-Ma (Eds.): SIMPDA 2015, LNBIP 244, pp. 142–154, 2017.
DOI: 10.1007/978-3-319-53435-0_7

is capable of processing large amounts of data in parallel and of quickly identifying patterns and outliers [34]. Thus, visualization helps to provide an overview of large and complex data. Furthermore, interactive controls allow the user to explore different aspects of the data in more detail.

Visual analytics is defined as 'the science of analytical reasoning facilitated by visual interactive interfaces' [45, p. 4]. It exploits both, the computational power of the computer and the human's perception system to facilitate insights and enable knowledge discovery in large and complex bodies of data. Thus, visual analytics combines both of the aforementioned approaches: machine learning and data mining, and visualization of the data.

1.1 Benefits of Visualization

One example that motivates the use of visualizations nicely, is the Anscombe's Quartet: In 1973 the statistician Francis Anscombe constructed four data sets [5] to demonstrate the necessity to analyze data with visual means as well as the impact of outliers on statistical properties. The summary statistic properties of these four data sets are nearly identical which could lead to the conclusion that these data sets are nearly identical. However, visualizing these data sets with scatter-plots reveals very different behavior of all four sets.

Another example is given by Spence and Garrison [44]: The Hertzsprung Russel Diagram shows the temperature of stars plotted against their magnitude. When looking at this diagram humans can easily distinguish different clusters of types of stars, while automatic means fail due to the noise and artifacts within the data. Card et al. summarize the ways is which information visualization amplifies human cognition like this [10]:

- Increasing working resources such as by using a visual resource to offload work from the cognitive to the perceptional system
- Reducing search such as by representing large amounts of data and grouping information
- Enhancing the recognition of patterns such as by visually organizing data by structural relationships
- Supporting perceptual inference of relationships by suiting the human's perceptual abilities
- Perceptual monitoring of a large number of potential events such as by making changes stand out visually
- Providing a manipulable medium that, unlike static diagrams, allows for interactively exploring the data

In contrast to pure question answering tools, such as IBM Watson [13] (a computing system that uses machine learning technologies to answer specific questions), visual analytics combines these technologies with interactive presentations of data and results, and thus, it enables users to derive additional insights from interacting with the data. Hence, users are able to browse the data and they may find answers to questions they did not even know they were looking for.

2   Visual Approaches for Process Mining

In their process mining manifesto [47] van der Aalst et al. emphasize the potential of visual analytics, i.e. the combination of visualizations, interactions, and mining techniques, to enhance process mining. We conducted a literature research starting with the process mining manifesto and references to relevant systems we already knew. Moreover, we used libraries and search engines such as Google Scholar [16] and IEEE Xplore [24] to search for various related keywords and we went through proceedings of important conferences such as IEEE VIS [23], EuroVis [12] conference, and the BPM conference [11], as well as through the list of references of relevant papers.

Most process mining approaches employ some kind of visual representation due to the wealth and complexity of the data. Yet, they usually do not explicitly consider visualization as a key factor and rather put emphasis on the mining methods [46,47]. Visualizations are used, for instance, for process discovery: to explore the event log data and to understand, reason about, and fine-tune the derived models (e.g. [20–22,26,32,48,51]). To this end, flow charts [36] and directed graphs are used most frequently. Other examples use visualizations to identify interesting behavior and patterns in the event log data (e.g., [7,43]), or to investigate process conformance (e.g. [1,8,49]). In this position paper I give by no means a comprehensive list of visualization approaches in the context of process mining. Yet, I introduce a selection of our own approaches, highlighting some alternative visual representations and interaction techniques in order to spark ideas how to combine process mining with visual analytics.

2.1   Visual Analysis of Guideline Conformance

A related topic to the analysis of process conformance is the analysis of conformance of clinical actions with the recommended care process. Clinical practice guidelines give specific recommendations on how to treat a patient in a given clinical situation. They are aimed at assisting the care process and ensuring its quality [14]. Hence, they can be seen as process models. Bodesinsky et al. [8] presented a visual interactive approach to analyze the compliance of clinical care with these guidelines. The approach is aimed at supporting (1) physicians at the point of care to display clinical guidelines and point them to omitted clinical actions, and (2) medical experts in retrospective analysis of the quality of the treatment and also of the quality of the clinical practice guideline. It visualizes the recommended sequence of actions (see Fig. 1(a)) together with the actual clinical actions that were applied (see Fig. 1(c)). Applied clinical actions are represented by diamonds on a time axis (a pop-up window gives the user details about these actions when hovering them). Moreover, the patient's parameters (see Fig. 1(b)) and the compliance of actions with the clinical guidelines (see Fig. 1(c)) are represented on a time axis. To aid the analysis of conformance, they visually distinguish conform and non-conform actions, as well as missing actions. Conform actions are indicated by diamonds and non-conform actions

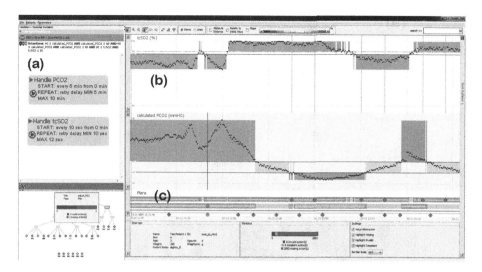

Fig. 1. Visual analysis of guideline conformance [8]. The recommended care process is visualized in a flow chart like graph in (a). In this screenshot the graph is zoomed in to show two repeated actions that should be carried out in parallel. (b) shows the patient's parameters that are affected by these actions. (c) represents actual clinical actions on a time axis. Actions that conform to the guideline are indicated by diamonds, non-conform actions are diamonds marked with an X, and the delay of actions is indicated by colored horizontal bars. A pop-up window gives the user details about these actions when hovering them. (Color figure online)

are indicated by diamonds marked with an X. Any delay of application is indicated by two parallel colored bars, marking the time span between the intended time of application and the actual time of application (see Fig. 1(c)).

2.2 Plan Strips

Another example from the medical domain is the Plan Strips visualization [39]. This interactive approach visualizes the complex hierarchies of processes – again using the example of clinical practice guidelines – by a set of nested rectangles. These guidelines recommend treatment plans (groups of clinical actions) which are composed of sub-plans. The execution order of a plan determines in which of the following ways its sub-plans are to be carried out: in sequence, in parallel (all sub-plans start at the same time), unordered (in sequence, in parallel, or a mixture of both), or in any-order (in sequence but the order is not defined) [40]. In addition, cyclical plans represent plans which are carried out in loops. Plan Strips encode these different kinds of sub-plan synchronization by color (see Fig. 2): Gray rectangles indicate recommended clinical actions or activities while colored containers represent the corresponding parent-plans as well as the execution order. Interactive selection and highlighting techniques are used to ease navigation. One limitation of this approach is that the visualization

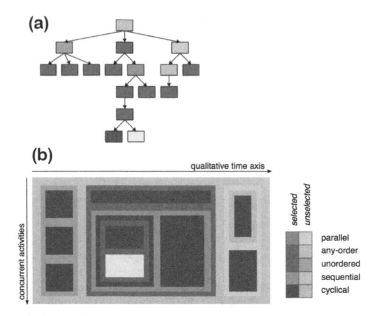

Fig. 2. Plan Strips [39] represent the hierarchical structure (a) of clinical plans and sub-plans as nested rectangles (b). User performed actions are indicated by gray rectangles (light gray when selected), while the color of parent containers indicate the execution order. (Color figure online)

demands for some learning effort, however, Plan Strips demonstrate an alternative way of representing highly specified and complex conditions of a process model in a space efficient way.

2.3 EventExplorer

EventExplorer [7] was designed to aid the analyst in exploring event log data, in order to gain some understanding of the data and the dependencies between events before deciding for concrete analysis methods or process mining algorithms. To this end, it combines pattern mining with interactive visualizations. Moreover, it supports the exploration of events on a temporal scale to aid the identification of important temporal dependencies of events (e.g., two events always occur together). Figure 3 shows a screenshot of EventExplorer. Each case (i.e., sequence of events) of the dataset is represented by a horizontal line composed of colored rectangles which represent the different events: The color indicates the type of event, while the order indicates the sequence of events. The user can switch the representation to lay out events along a time axis to investigate their precise temporal behavior. EventExplorer uses pattern mining techniques to identify patterns within sequences of events. When a pattern is selected, all instances of this pattern within one case are highlighted and connected by gray arcs which emphasize the recurrence of this pattern within

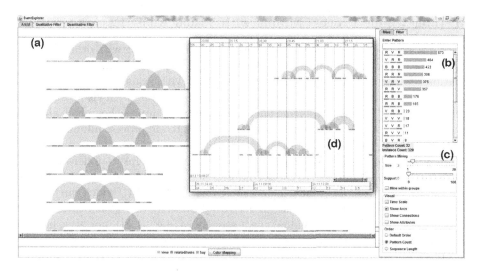

Fig. 3. EventExplorer [7]. In (a) EventExplorer represents all cases of the dataset as horizontal lines with color coded events (scrollable). (b) shows the different patterns (together with their number of occurrences) which match the settings in (c), i.e., the length of the pattern and the minimum number of occurrences. Alternatively, the user can enter a specific pattern of interest into the text field in (b). Selected patterns (in this case 'V-R-V') are highlighted by gray arcs in (a). The user can switch the x-axis in (a) from a sequential axis to a time axis (d) in order to explore the precise temporal characteristics of event sequences. (Color figure online)

the case. EventExplorer also supports fuzzy pattern mining by allowing the use of wildcards when entering a pattern of interest. The approach presents interactive visualizations to (1) get familiar with the data, (2) to cluster and sort cases, and (3) to identify temporal patterns. The arcs visually emphasize recurring patterns as well as temporal distances (when using the time axis view).

The three approaches outlined in detail were designed for very diverse purposes and give only examples of applications of visual analytics in process mining. However, there are still many open challenges but also opportunities when looking at the combination of these two research fields. We derived these challenges from the literature review, our previous work, and our long lasting experience in the research field of visual analytics.

3   Challenges and Opportunities

Researchers have identified challenges of today's process mining research in various works (e.g., [33,47]). In [47] van der Aalst et al. even mention the potential of visual analytics to support process mining. Others again have summarized open challenges in the field of visual analytics (e.g., [30,45,50]). When focusing on the combination of visual analytics and process mining, we face related challenges.

C1: Intertwining Process Mining with Visual Analytics

Van der Aalst et al. identify the combination of process mining with other types of analysis as one of the challenges in process mining [47]. Moreover, they explicitly mention the potential of process mining combined with visual analytics for leading to further insights. Yet, how to intertwine process mining with visual analytics techniques is still an open problem. There are a couple of approaches which already tackle this problem. Among these visual approaches to aid process mining, the use of flow charts [36] or directed graphs to represent the process model are most common (e.g., [20–22,32,39]). Visualizing the process model helps to understand and fine-tune the model, but also to check conformance. Other visual approaches show an overview of individual event sequences (e.g., [7,26,43,48,51]) which can be used to identify common paths and thus, to derive a process model. Visualizations that show the conformance of individual event sequences (e.g., [1,8]), however, help to identify shortcomings of the existing model, but also variations of individual cases. Yet, there is still much unexploited potential of visual analytics to better support pattern discovery and the derivation of process models, fine-tuning and enhancing these models, as well as identifying conformance problems and finding alternative solutions. To this end, visualizations need to be tightly coupled to analytical techniques and mining algorithms in an iterative loop, giving immediate visual feedback to changes and adjustments made to the algorithms. Appropriate views and interactions need to be tailored to specific user groups with respect to their specific tasks and data [35].

C2: Scalability and Aggregation

The scalability of visualizations is a well-known challenge in visual analytics [30,45,50]. This is also true in the context of process mining, which usually deals with huge amounts of event sequences that need to be analyzed. On the one hand, process mining algorithms tackle this problem by reducing these huge event logs to a manageable amount of patterns and process models. On the other hand, the analysis of the raw data and single event sequences may yield some important insights, and thus, visualizations need to support the simultaneous analysis of huge amounts of individual event sequences. There are some approaches that tackle this problem to some extent (e.g., [7,26,51]) by simultaneously showing a number of individual event sequences at minimum display space. Still the number of individual event sequences that can be displayed simultaneously is limited and it is an open challenge how to design interactive visualizations to support the whole analysis cycle from single event sequences to multiple event logs. There is a need for scalable visualizations able to represent a big number of cases while allowing zoom-in and drill-down to investigate details about single event sequences, in analogy to the famous Information Seeking Mantra by Shneiderman: 'overview first, zoom and filter, then details-on-demand' [41, p. 337].

C3: Interaction to Support Process Discovery and Enhancement

This challenge is tightly coupled with the last challenge of finding appropriate visualization techniques to support the different tasks. Data exploration is an iterative process with trial and error loops. Suitable interaction techniques need to support drilling down from an overview visualization to the investigation of single events. Commonly used techniques include zoom and filter functionalities, however, other interaction techniques may be needed depending on the data, user, and task [35]. Further interaction techniques include *select, explore, reconfigure, encode, abstract/elaborate,* and *connect* [52]. Allowing the analyst to interactively investigate the raw event data as well as the effects of changes made to models and mining algorithms may foster new insights and improve process mining results.

EventExplorer [7] allows for sorting cases by pattern count or by time. Another interesting way of sorting would be by similarity of event sequences in order to aid the identification of different groups of cases. Moreover, it supports the investigation of events in sequence or on a temporal scale. An interesting addition would be allowing the analyst to apply different ways of aligning cases vertically, for instance, aligning cases with the first appearance of a specific event.

C4: Data Quality and Uncertainty

Data quality is a general challenge in visual analytics [27], as well as in process mining [47], and thus, data quality is just as important when it comes to the combination of both fields. While process mining algorithms require good data quality and well-structured event logs, in practice these can be erroneous and badly structured. Instances of the same event type may have different names, event granularities may vary, and quite often the event log contains missing, incorrect, imprecise, uncertain, or irrelevant data. Data quality control can be divided into (1) data profiling (i.e., identifying and communicating quality problems), (2) data cleansing (i.e., correcting erroneous data), and (3) data transformation (i.e., transforming the data into appropriate formats for automatic processing) [19,27]. There are visual approaches that tackle the problem of data quality (e.g., [9,17,28,29]), however, the specific needs in process mining are hardly considered, such as case heterogeneity, event granularity, or concept drift.

On the other hand, data often contains some amount of uncertainty. Event log data may, for instance, contain uncertainties about which event type corresponds to which log entry or about the exact time of an event. Visually communicating these uncertainties to the analyst (e.g., [18]) is important for a better informed reasoning.

C5: Complexity of Time-Oriented Data

Usually event log data has an inherent temporal structure. For any such event log at least the sequence of events is known and used to analyze paths and processes. Yet, there are also event logs with precise timestamps for each event, allowing for a detailed analysis regarding the temporal dimension of events and event sequences. Time, however, is a complex data domain with very special features that require special consideration [2,4,41]. Some examples of these special characteristics are: time can be given as time points or intervals, conventionally time is aggregated irregularly (i.e., 60 min per hour, 24 h per day, 28–31 days per month, 356/366 days per year, ...), leap seconds and days, different time zones, recurrences (e.g., seasonal cycles), and it has a strong social component (e.g., weekends, holidays) [2]. Ailenei et al. [3] outline use cases with special consideration of the temporal behavior of processes that need to be considered in process mining. When it comes to the visualization of time-oriented data, Aigner et al. [2] give a comprehensive overview. However, considering these peculiar characteristics of events (e.g., type of day, some cyclic behavior synchronized with the calendar, or events that co-occur with a certain delay) and adequately visualizing them for the task at hand is still an open challenge.

C6: Evaluation

Since visual analytics aims at supporting human reasoning and gaining new insights, the quality of visual analytics approaches is hard to quantify. Plaisant [37] outlines different challenges of evaluating visualizations and emphasizes that evaluation strategies need to take the exploratory nature of tasks and the added value of visualization such as overall awareness and potential discoveries into account. Moreover, praxis related aspects such as a successful adoption of these visualizations need to be considered. There are different approaches for evaluating visualizations [15,25,42] which usually involve target user participation and are highly task dependent (see [4] for a task framework and [31] for a time-specific task framework). In any case it is recommended to pursue an iterative design approach which involves feedback of the target user as early as possible.

On the other hand, evaluation in process mining is an open challenge too. There are different approaches on how to evaluate derived process models (e.g., [6,38]), and Ailenei et al. [3] present a set of task specific use cases for the evaluation of process mining tools. These use cases may as well help to evaluate visual analytics solutions for process mining from a technical point of view. However, the evaluation of gained awareness and insights is still an open challenge.

4   Conclusion

Handling huge amounts of process data can be tackled in different ways: automatic mining techniques, visualizations, or by visual analytics which combines the benefits of both fields. Especially the latter has potential to improve the results of process analysis in many ways. In this paper I highlighted some examples of how this was approached in recent years. Moreover, I outlined six open challenges which reveal opportunities for further research. I believe, visual analytics is beneficial to many tasks of process mining and – if done right – will have substantial impact on the quality of process analysis.

Acknowledgements. The research leading to these results has received funding from the Centre for Visual Analytics Science and Technology CVAST, funded by the Austrian Federal Ministry of Science, Research, and Economy in the exceptional Laura Bassi Centres of Excellence initiative (#822746).

References

1. Adriansyah, A., van Dongen, B., van der Aalst, W.M.P.: Conformance checking using cost-based fitness analysis. In: Proceedings of the Enterprise Computing Conference (EDOC), pp. 55–64, Piscataway, NJ, USA. IEEE Educational Activities Department (2011)
2. Aigner, W., Miksch, S., Schuman, H., Tominski, C.: Visualization of Time-Oriented Data. Human-Computer Interaction, 1st edn. Springer, Heidelberg (2011)
3. Ailenei, I., Rozinat, A., Eckert, A., van der Aalst, W.M.P.: Definition and validation of process mining use cases. In: Daniel, F., Barkaoui, K., Dustdar, S. (eds.) BPM 2011. LNBIP, vol. 99, pp. 75–86. Springer, Heidelberg (2012). doi:10.1007/978-3-642-28108-2_7
4. Andrienko, N., Andrienko, G.: Exploratory Analysis of Spatial and Temporal Data: A Systematic Approach. Springer, Heidelberg (2006)
5. Anscombe, F.J.: Graphs in statistical analysis. Am. Stat. **27**(1), 17–21 (1973)
6. Huang, Z., Kumar, A.: New quality metrics for evaluating process models. In: Ardagna, D., Mecella, M., Yang, J. (eds.) BPM 2008. LNBIP, vol. 17, pp. 164–170. Springer, Heidelberg (2009). doi:10.1007/978-3-642-00328-8_16
7. Bodesinsky, P., Alsallakh, B., Gschwandtner, T., Miksch, S.: Exploration and assessment of event data. In: Bertini, E., Roberts, J.C. (eds.) Sixth International EuroVis Workshop on Visual Analytics (EuroVA) 2015, p. 5. The Eurographics Association (2015)
8. Bodesinsky, P., Federico, P., Miksch, S.: Visual analysis of compliance with clinical guidelines. In: Proceedings of the 13th International Conference on Knowledge Management and Knowledge Technologies (i-KNOW 2013), pp. 12:1–12:8. ACM (2013)
9. Bors, C., Gschwandtner, T., Miksch, S., Gärtner, J.: Qualitytrails: data quality provenance as a basis for sensemaking. In: Xu, K., Attfield, S., Jankun-Kelly, T.J. (eds.) Proceedings of the IEEE VIS Workshop on Provenance for Sense-making, pp. 1–2 (2014)
10. Card, S.K., Mackinlay, J.D., Shneiderman, B. (eds.): Readings in Information Visualization: Using Vision to Think. Morgan Kaufmann Publishers Inc., San Francisco (1999)
11. Conferences on Business Process Management. http://bpm-conference.org/BpmConference/. Accessed 25 Aug 2016
12. EuroVis (EG/VGTC Conference on Visualization). https://www.eg.org/index.php/about-eg/working-groups/67-about-eg/working-groups/273-working-group-on-data-visualization-events. Accessed 25 Aug 2016
13. Ferrucci, D.A.: Introduction to "This is Watson". IBM J. Res. Dev. **56**(3.4), 1–15 (2012)

14. Field, M.J., Lohr, K.N. (eds.): Clinical Practice Guidelines: Directions for a New Program. National Academies Press, Institute of Medicine, Washington DC (1990). http://www.nap.edu/books/0309043468/html/. Accessed Apr 2016
15. Forsell, C., Cooper, M.: An introduction and guide to evaluation of visualization techniques through user studies. In: Huang, W. (ed.) Handbook of Human Centric Visualization, pp. 285–313. Springer, New York (2014)
16. Google Scholar. https://scholar.google.at/. Accessed 25 Aug 2016
17. Gschwandtner, T., Aigner, W., Miksch, S., Gärtner, J., Kriglstein, S., Pohl, M., Suchy, N.: TimeCleanser: a visual analytics approach for data cleansing of time-oriented data. In: Lindstaedt, S., Granitzer, M., Sack, H. (eds.) 14th International Conference on Knowledge Technologies and Data-driven Business (i-KNOW 2014), pp. 1–8. ACM Press (2014)
18. Gschwandtner, T., Bögl, M., Federico, P., Miksch, S.: Visual encodings of temporal uncertainty: a comparative user study. IEEE Trans. Vis. Comput. Graph. **22**, 539–548 (2016)
19. Gschwandtner, T., Gärtner, J., Aigner, W., Miksch, S.: A taxonomy of dirty time-oriented data. In: Quirchmayr, G., Basl, J., You, I., Xu, L., Weippl, E. (eds.) CD-ARES 2012. LNCS, vol. 7465, pp. 58–72. Springer, Heidelberg (2012). doi:10. 1007/978-3-642-32498-7_5
20. Günther, C.W., van der Aalst, W.M.P.: Fuzzy mining – adaptive process simplification based on multi-perspective metrics. In: Alonso, G., Dadam, P., Rosemann, M. (eds.) BPM 2007. LNCS, vol. 4714, pp. 328–343. Springer, Heidelberg (2007). doi:10.1007/978-3-540-75183-0_24
21. Günther, C.W., Rozinat, A.: Disco: discover your processes. In: Proceedings of the Demonstration Track of the 10th International Conference on Business Process Management (BPM 2012), Tallinn, Estonia, September 4, 2012, pp. 40–44 (2012)
22. Hipp, M., Mutschler, B., Reichert, M.: Navigating in process model collections: a new approach inspired by Google earth. In: Daniel, F., Barkaoui, K., Dustdar, S. (eds.) BPM 2011. LNBIP, vol. 100, pp. 87–98. Springer, Heidelberg (2012). doi:10. 1007/978-3-642-28115-0_9
23. IEEE VIS. http://ieeevis.org/. Accessed 25 Aug 2016
24. IEEE Xplore Digital Library. http://ieeexplore.ieee.org/Xplore/home.jsp. Accessed 25 Aug 2016
25. Isenberg, T., Isenberg, P., Chen, J., Sedlmair, M., Möller, T.: A systematic review on the practice of evaluating visualization. IEEE Trans. Vis. Comput. Graph. **19**(12), 2818–2827 (2013)
26. Jagadeesh, R.P., Bose, C., van der Aalst, W.M.P.: Process diagnostics using trace alignment: opportunities, issues, and challenges. Inf. Syst. **37**(2), 117–141 (2012)
27. Kandel, S., Heer, J., Plaisant, C., Kennedy, J., van Ham, F., Riche, N.H., Weaver, C., Lee, B., Brodbeck, D., Buono, P.: Research directions in data wrangling: visualizations and transformations for usable and credible data. Inf. Vis. **10**(4), 271–288 (2011)
28. Kandel, S., Paepcke, A., Hellerstein, J., Heer, J.: Wrangler: interactive visual specification of data transformation scripts. In: Proceedings of the ACM Conference Human Factors in Computing Systems (CHI 2011), pp. 3363–3372, May 2011
29. Kandel, S., Parikh, R., Paepcke, A., Hellerstein, J., Heer, J.: Profiler: integrated statistical analysis and visualization for data quality assessment. In: Proceedings of the International Working Conference on Advanced Visual Interfaces (AVI 2012), pp. 547–554, May 2012

30. Keim, D., Andrienko, G., Fekete, J.-D., Görg, C., Kohlhammer, J., Melançon, G.:
    Visual analytics: definition, process, and challenges. In: Kerren, A., Stasko, J.T.,
    Fekete, J.-D., North, C. (eds.) Information Visualization. LNCS, vol. 4950, pp.
    154–175. Springer, Heidelberg (2008). doi:10.1007/978-3-540-70956-5_7
31. Lammarsch, T., Rind, A., Aigner, W., Miksch, S.: Developing an extended task
    framework for exploratory data analysis along the structure of time. In: Matkovic,
    K., Santucci, G. (eds.) Proceedings of the EuroVis Workshop on Visual Analytics
    in Vienna, Austria (EuroVA 2012), pp. 31–35. Eurographics (2012)
32. Maguire, E., Rocca-Serra, P., Sansone, S., Davies, J., Chen, M.: Visual compression
    of workflow visualizations with automated detection of macro motifs. IEEE Trans.
    Vis. Comput. Graph. **19**(12), 2576–2585 (2013)
33. Mans, R.S., van der Aalst, W.M.P., Vanwersch, R.J.B., Moleman, A.J.: Process
    mining in healthcare: data challenges when answering frequently posed questions.
    In: Lenz, R., Miksch, S., Peleg, M., Reichert, M., Riaño, D., Teije, A. (eds.)
    KR4HC/ProHealth -2012. LNCS (LNAI), vol. 7738, pp. 140–153. Springer, Hei-
    delberg (2013). doi:10.1007/978-3-642-36438-9_10
34. Mazza, R.: Perception. In: Mazza, R. (ed.) Introduction to Information Visualiza-
    tion, Chapter 3, pp. 33–44. Springer, Heidelberg (2009)
35. Miksch, S., Aigner, W.: A matter of time: applying a data-users-tasks design tri-
    angle to visual analytics of time-oriented data. Comput. Graph. Spec. Sect. Vis.
    Anal. **38**, 286–290 (2014)
36. Peters, K.: http://home.southernct.edu/~petersk1/csc400/csc400-owchart.htm.
    Accessed 05 Jan 2009
37. Plaisant, C.: The challenge of information visualization evaluation. In: Proceedings
    of the Working Conference on Advanced Visual Interfaces (AVI 2004), pp. 109–116.
    ACM, New York (2004)
38. Rozinat, A., de Medeiros, A.K.A., Günther, C.W., Weijters, A.J.M.M., van der
    Aalst, W.M.P.: The need for a process mining evaluation framework in research and
    practice. In: Hofstede, A., Benatallah, B., Paik, H.-Y. (eds.) BPM 2007. LNCS, vol.
    4928, pp. 84–89. Springer, Heidelberg (2008). doi:10.1007/978-3-540-78238-4_10
39. Seyfang, A., Kaiser, K., Gschwandtner, T., Miksch, S.: Visualizing complex process
    hierarchies during the modeling process. In: Rosa, M., Soffer, P. (eds.) BPM
    2012. LNBIP, vol. 132, pp. 768–779. Springer, Heidelberg (2013). doi:10.1007/
    978-3-642-36285-9_77
40. Shahar, Y., Miksch, S., Johnson, P.: The Asgaard Project: a task-specific frame-
    work for the application and critiquing of time-oriented clinical guidelines. Artif.
    Intell. Med. **14**, 29–51 (1998)
41. Shneiderman, B.: The eyes have it: a task by data type taxonomy for information
    visualizations. In: Proceedings of the 1996 IEEE Symposium on Visual Languages,
    Piscataway, NJ, USA, September 1996, pp. 336–343. IEEE Educational Activities
    Department (1996)
42. Shneiderman, B., Plaisant, C.: Strategies for evaluating information visualization
    tools: multi-dimensional in-depth long-term case studies. In: Proceedings of the
    2006 AVI Workshop on Beyond Time and Errors: Novel Evaluation Methods for
    Information Visualization, BELIV 2006, pp. 1–7. ACM, New York (2006)
43. Song, M., van der Aalst, W.M.P.: Supporting process mining by showing events at a
    glance. In: Proceedings of the 17th Annual Workshop on Information Technologies
    and Systems (WITS), pp. 139–145 (2007)
44. Spence, I., Garrison, R.F.: A remarkable scatterplot. Am. Stat. **47**, 12–19 (1993)

45. Thomas, J.J., Cook, K.A. (eds.) Illuminating the path: the research and development agenda for visual analytics. IEEE Educational Activities Department, Piscataway (2005)

46. van der Aalst, W.M.P. (ed.): Process Mining: Discovery Conformance and Enhancement of Business Processes, 1st edn. Springer, Heidelberg (2011)

47. van der Aalst, W.M.P., et al.: Process mining manifesto. In: Daniel, F., Barkaoui, K., Dustdar, S. (eds.) BPM 2011. LNBIP, vol. 99, pp. 169–194. Springer, Heidelberg (2012). doi:10.1007/978-3-642-28108-2_19

48. Vrotsou, K., Johansson, J., Cooper, M.: Activitree: interactive visual exploration of sequences in event-based data using graph similarity. IEEE Trans. Vis. Comput. Graph. 15(6), 945–952 (2009)

49. Wikipedia. http://en.wikipedia.org/wiki/gantt_chart. Accessed 22 Dec 2011

50. Wong, P.C., Shen, H., Johnson, C.R., Chen, C., Ross, R.B.: The top 10 challenges in extreme-scale visual analytics. IEEE Comput. Graph. Appl. 32(4), 63–67 (2012)

51. Wongsuphasawat, K., Gotz, D.: Exploring flow, factors, and outcomes of temporal event sequences with the outflow visualization. IEEE Trans. Vis. Comput. Graph. 18(12), 2659–2668 (2012)

52. Yi, J.S., ah Kang, Y., Stasko, J.T., Jacko, J.A.: Toward a deeper understanding of the role of interaction in information visualization. IEEE Trans. Vis. Comput. Graph. 13(6), 1224–1231 (2007)

A Relational Data Warehouse
for Multidimensional Process Mining

Thomas Vogelgesang$^{(\boxtimes)}$ and H.-Jürgen Appelrath

Department of Computer Science, University of Oldenburg, Oldenburg, Germany
thomas.vogelgesang@uni-oldenburg.de

Abstract. Multidimensional process mining adopts the concept of data cubes to split event data into a set of homogenous sublogs according to case and event attributes. For each sublog, a separated process model is discovered and compared to other models to identify group-specific differences for the process. For an effective explorative process analysis, performance is vital due to the explorative characteristics of the analysis. We propose to adopt well-established approaches from the data warehouse domain based on relational databases to provide acceptable performance. In this paper, we present the underlying relational concepts of PMCube, a data-warehouse-based approach for multidimensional process mining. Based on a relational database schema, we introduce generic query patterns which map OLAP queries onto SQL to push the operations (i.e. aggregation and filtering) to the database management system. We evaluate the run-time behavior of our approach by a number of experiments. The results show that our approach provides a significantly better performance than the state-of-the-art for multidimensional process mining and scales up linearly with the number of events.

1 Introduction

Process mining comprises a set of techniques that allows for the automatic analysis of (business) processes. It is based on so-called event logs which consist of events recorded during the execution of the process. Figure 1 illustrates the typical structure of event logs. The events are grouped by their respective process instance (case) and the ordered sequence of events for a case forms the trace. Both, cases and events, may store arbitrary information as attributes.

According to van der Aalst [13], there are three different kinds of process mining: (1) Process discovery extracts a process model from the event log reflecting its behavior, (2) conformance checking compares an event log to a manually created or previously discovered process model to measure the quality of the process model, and (3) process enhancement extends a process model with additional information (e.g., time-stamps) to provide additional perspectives on the process.

Besides traditional business processes, process mining can also be applied in the domain of healthcare e.g., to analyze the treatment process in a hospital.

P. Ceravolo and S. Rinderle-Ma (Eds.): SIMPDA 2015, LNBIP 244, pp. 155–184, 2017.
DOI: 10.1007/978-3-319-53435-0_8

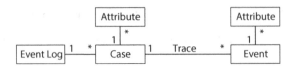

Fig. 1. General structure of an event log

However, healthcare processes are typically unstructured due to the individuality of patients. The treatment has to be adjusted to the individual situation of the patient considering age, sex, other diseases, and other features of the patient. Furthermore, the process may also be influenced by institutional features, e.g., the experience of the medical staff. To minimize the influence of such features, it is desirable to group patients with similar features and to analyze the process separately for each group. Otherwise, the heterogeneity of patients would result in a very complex model blurring the influence of particular features.

Traditional process mining techniques only consider the entire event log. Even though filters can be applied to restrict the event log to a particular subset of cases or events, this requires high effort if multiple groups of patients should be analyzed and compared to each other. Therefore, an approach is required that enables the analyst to partition event logs into groups of cases with homogeneous features in a dynamic and flexible way. Then, an individual process model for each group can be separately mined and compared to other models.

Multidimensional process mining (MPM) achieves this by adopting the concept of data cubes that is well-known in the domain of data warehousing. It considers the attributes of the event log, describing the variable features of the patients, as dimensions forming a multidimensional data space. Each combination of dimension values forms a cell of the cube that contains a subset of the event log (sublog) related to these dimension values. OLAP (Online Analytical Processing) operators [3] can be used to manipulate the data cube and define specific views on the data. For instance, roll-up and drill-down operators can be applied to change the granularity of the cube's dimensions while slice and dice operators can be used to filter the data.

MPM is characterized by its explorative approach. The OLAP queries are gradually modified to analyze the processes from multiple views. This way, the analyst can derive and verify new hypothesis. To avoid interruptions of the explorative workflow, it is important to keep waiting times as short as possible. Therefore, performance is vital for MPM, even though it is not a time-critical application. We propose to adopt well-established data warehouse (DWH) technologies based on relational databases, to provide satisfying loading times for the multidimensional event data.

In this paper, we show how to link multidimensional event data to relational databases for MPM. As our main contribution, we show how to express the MPM-specific OLAP queries using SQL to push filtering and aggregation of event data into the database management system (DBMS). This way, MPM may benefit from the comprehensive optimization techniques provided by state-of-the-art DBMS.

The paper is organized as follows. First, we discuss related work in Sect. 2 and introduce a running example in Sect. 3. The general concept of our approach PMCube is briefly introduced in Sect. 4. While Sect. 5 presents the logical data model of the underlying data warehouse, Sect. 6 explains its mapping onto a relational database schema. In Sect. 7, we map high-level OLAP operators onto generic patterns expressed in SQL. In Sect. 8, we evaluate our approach by a number of performance measurements comparing our approach to the state-of-the-art approach for MPM. We conclude our paper in Sect. 9.

2 Related Work

There is a wide range of literature in the data warehousing domain (e.g., [3]) describing the different general approaches for the implementation of data cubes. Multidimensional OLAP (MOLAP) approaches rely on a mainly memory-based multidimensional array storage. Relational OLAP (ROLAP) maps the multi-dimensional data onto a relational database schema. A combination of both approaches is known as Hybrid OLAP (HOLAP). In ROLAP approaches, the schema typically consists of a fact table storing the data values. This table is linked to other tables storing the values of the dimension and their hierarchies. In the star schema, each dimension is stored in a single table representing all its dimension levels while the snowflake schema stores each dimension level in its own table.

The process mining manifesto [15] gives an overview of the field of process mining. For a comprehensive introduction to this topic, we refer to van der Aalst [13]. Event Cubes [11] are a first approach for MPM. This approach uses information retrieval techniques to create an index over a traditionally structured event log and derives a data cube from it. Each cell of an event cube contains precomputed dependency measures instead of raw event data. A single process model is generated on-the-fly from these dependency measures where each value of the considered dimensions is mapped onto a different path in the model.

Process Cubes [14] are another approach for MPM which uses OLAP operators to partition the event data into sublogs. It combines all dimensions in a common data space so the cells of the cube contain sets of events. Its implementation Process Mining Cube (PMC) [1] can use different algorithms to discover a process model for each cell and provides the visualization of multiple process models in a grid. Both, Event Cubes and PMC, are based on a MOLAP approach. According to Bolt and Aalst [1], PMC provides a significantly better performance than the previous Process Cubes implementation PROCUBE. However, the reported loading times are still quite long if related to the amount of data. Additionally, the filtering is limited to filtering particular cases and events. Moreover, it does not provide the aggregation of events into high-level events.

Besides, there are several approaches for special DWH for storing process data (e.g. [8,9]). These process warehouses (PWH) aim to analyze the underlying processes to identify problems in process execution. However, they do not store complete event logs, but measures for process performance (e.g. execution times), where events and cases form the dimensions of the cube. The analysis is performed by aggregating the measures along the dimensions. In contrast to MPM, these approaches generally do not support process mining.

Relational XES [16] is a data format for storing event logs which is based on the XES [4] event log structure. In contrast to it, Relational XES uses a relational database in the background to store the data in a relational structure. In comparison to the usual XML-based XES format, Relational XES reduces the overhead for storing big event logs. However, it does not provide a multidimensional structure to store the event logs in a data cube to directly support MPM.

Schönig et al. present an approach for declarative process mining [12] that assumes the data to be stored in a database using the Relational XES schema. In order to use the optimization concepts provided by the database, they define query patterns in SQL to extract the constraints of the declarative process model directly from the database. In contrast to our approach, they perform process discovery directly in the database system which is however limited to declarative process models. Furthermore, they do not either consider multidimensional aspects of the data nor compare different variants of the same process.

An approach for data-aware declarative process mining is described in [7]. It extends the logical constraints of the declarative process model notation by additional conditions defined over the data attributes. Roughly, this approach is comparable to process discovery with an integrated analysis of decision points. Even though it considers arbitrary data attributes of the event logs, it does not provide a multidimensional view on the processes. In contrast to the most approaches for MPM, it generates a single process model instead of a set of variant-specific models.

The approach presented in [10] aims to automatically analyze the structure of relational database items like tables, primary keys, and foreign key relationships to identify so-called business objects or artifacts. Based on the identified artifacts, it automatically generates SQL queries to extract event logs from the database that reflect different processes and are suitable for process mining. However, this approach does not consider MPM.

The Slice-Mine-and-Dice (SMD) approach [2] aims to reduce the complexity of process models in two ways. On the one hand, it clusters similar traces of the event log to split the model into a set of behavioral variants. On the other hand, it reduces the model complexity by extracting and merging of shared subprocesses. Even though the name of the approach indicates a relationship to MPM, it neither considers attribute values nor it uses data cubes or OLAP operations.

3   Running Example

To provide a descriptive explanation of the developed concepts, we introduce a running example in this section. We assume a simple and fictive healthcare

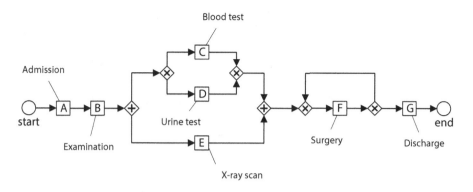

Fig. 2. Simple and fictive healthcare process of the running example

process which comprises the diagnoses and therapy for a patient. Figure 2 shows this process as a BPMN model.

The process starts with the admission of the patient to the hospital (A). After the examination of the patient (B), the process splits into two parallel paths. The first path consists of a lab test, which can be either a blood test (C) or an urine test (D). The second path consists of a X-ray scan (E). After both parallel paths are completed, the treatment proceeds with a surgery (F) which can be repeated multiple times, e.g., in case of complications. Finally, the process ends after the patient's discharge from hospital (G).

For the running example, we define each hospital stay as an independent process instance. Therefore, a particular patient can be represented by several cases. We assume that each case is recorded in an event log with additional information. Age, sex, and insurance status of the patients, year of treatment, and the hospital are saved for the cases. Besides the activity of each event, it is also stored who performed the activity (a doctor or a nurse) and the costs related to the execution of the activity. Furthermore, we assume that each event has a time-stamp which is precise enough to reflect a unique event order.

Table 1 shows an example event log consisting of 20 cases reflecting the fictive example process of Fig. 2. For each case, its id and the age, the sex, the insurance status (public or private insurance), the year of treatment, and the hospital location are given. Furthermore, the table shows the trace of each case as the sequence of events using the abbreviations of activity names introduced in Fig. 2, e.g., ABCEFFG for the first case. In total, the event log contains four different traces representing different process behavior (variations are underlined): ABCEFG, ABEDFG, ABCEFFG, ABEDFFG. For the sake of clarity, we omit the event attributes.

Even though the process model (Fig. 2) reflects the overall process, it does not reveal the influence of particular attributes. Consequently, the specific process behavior of particular groups of patients cannot be seen in the model. Partitioning the event log L_1 into sublogs using MPM enables the analyst to identify differences in the group-specific process models that are related to particular

Table 1. Example event log L_1

Case id	Age	Sex	Insurance	Year	Hospital	Trace
1	34	Female	Private	2012	Oldenburg	ABCEFFG
2	31	Male	Private	2013	Vienna	ABCEFFG
3	23	Female	Public	2014	Oldenburg	ABCEFG
4	56	Male	Public	2012	Oldenburg	ABEDFG
5	69	Female	Public	2015	Vienna	ABEDFG
6	56	Male	Public	2014	Vienna	ABEDFG
7	54	Female	Private	2012	Vienna	ABEDFFG
8	55	Male	Public	2015	Oldenburg	ABEDFG
9	36	Female	Public	2015	Vienna	ABCEFG
10	67	Male	Public	2014	Oldenburg	ABEDFG
11	56	Female	Private	2013	Oldenburg	ABEDFFG
12	68	Male	Private	2014	Berlin	ABEDFFG
13	30	Female	Public	2014	Vienna	ABCEFG
14	71	Male	Public	2012	Berlin	ABEDFG
15	65	Female	Public	2015	Berlin	ABEDFG
16	23	Male	Private	2012	Vienna	ABCEFFG
17	29	Female	Public	2011	Vienna	ABCEFG
18	72	Male	Private	2010	Oldenburg	ABEDFFG
19	38	Female	Public	2010	Vienna	ABCEFG
20	42	Male	Public	2011	Berlin	ABCEFG

attributes and their values. Tables 2 and 3 show the two sublogs L_2 and L_3 which can be derived from L_1 using MPM by drilling-down the data cube alongside the age dimension. While L_2 comprises all cases of younger patients (age < 50), L_3 only contains the cases of older patients. For the sake of clarity, we only show the age, sex, and insurance status attributes.

Mining both sublogs results in the two variants of the process model which are shown in Fig. 3a (for L_2) and in Fig. 3b (for L_3). Comparing these models clearly shows that patients are treated differently depending on their age: The activity C (blood test) is only performed for patients younger than 50 years, the activity D (urine test) is exclusively performed for patients of 50 years or older.

Alternatively, drilling down alongside the insurance status dimension instead of the age dimension would reveal that activity F (surgery) is only repeated for patients with private healthcare insurance. For patients with public healthcare insurance, this activity is only executed once. However, partitioning the event log does not always result in such obvious differences. For instance, the sex of the patients does not have any influence on the treatment of the example process. Using the sex of the patients for drilling down results in the two sublogs L_4 (female) and L_5 (male) which are shown in Tables 4 and 5, respectively. Both

Table 2. Sublog L_2 (age < 50)

Case id	Age	Sex	Ins.	Trace
1	34	Female	Private	ABCEFFG
2	31	Male	Private	ABCEFFG
3	23	Female	Public	ABCEFG
9	36	Female	Public	ABCEFG
13	30	Female	Public	ABCEFG
16	23	Male	Private	ABCEFFG
17	29	Female	Public	ABCEFG
19	38	Female	Public	ABCEFG
20	42	Male	Public	ABCEFG

Table 3. Sublog L_3 (age ≥ 50)

Case id	Age	Sex	Ins.	Trace
4	56	Male	Public	ABEDFG
5	69	Female	Public	ABEDFG
6	56	Male	Public	ABEDFG
7	54	Female	Private	ABEDFFG
8	55	Male	Public	ABEDFG
10	67	Male	Public	ABEDFG
11	56	Female	Private	ABEDFFG
12	68	Male	Private	ABEDFFG
14	71	Male	Public	ABEDFG
15	65	Female	Public	ABEDFG
18	72	Male	Private	ABEDFFG

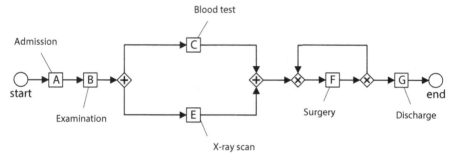

(a) Process model variant for sublog L_2 (age < 50)

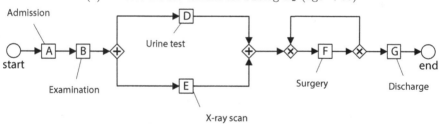

(b) Process model variant for sublog L_3 (age ≥ 50)

Fig. 3. Process model variants for sublogs L_2 and L_3

sublogs comprise all four variations of the process. Consequently, mining the sublogs L_4 and L_5 will result in the overall process model shown in Fig. 2.

In contrast to the given example, the influence of the dimensions is usually not that clear for real-world data. Even if a dimension has an affect on the process, the sublogs may contain noise due to exceptional behavior. The influence of a

Table 4. Sublog L_4 (female patients)

Case id	Age	Sex	Ins.	Trace
1	34	Female	Private	ABCEFFG
3	23	Female	Public	ABCEFG
5	69	Female	Public	ABEDFG
7	54	Female	Private	ABEDFFG
9	36	Female	Public	ABCEFG
11	56	Female	Private	ABEDFFG
13	30	Female	Public	ABCEFG
15	65	Female	Public	ABEDFG
17	29	Female	Public	ABCEFG
19	38	Female	Public	ABCEFG

Table 5. Sublog L_5 (male patients)

Case id	Age	Sex	Ins.	Trace
2	31	Male	Private	ABCEFFG
4	56	Male	Public	ABEDFG
6	56	Male	Public	ABEDFG
8	55	Male	Public	ABEDFG
10	67	Male	Public	ABEDFG
12	68	Male	Private	ABEDFFG
14	71	Male	Public	ABEDFG
16	23	Male	Private	ABCEFFG
18	72	Male	Private	ABEDFFG
20	42	Male	Public	ABCEFG

dimension may also not be distinctive. This means that the dimension value does not necessarily imply an exclusive choice like always executing activity C (blood test) for younger patients and activity D (urine test) for older patients, as it was shown in the example. Instead, the dimensions often influence the likelihood for a particular path in the process. Furthermore, process variations may only occur for a specific value combination of multiple dimensions.

4   PMCube Concept

Figure 4 illustrates the basic concept of PMCube. The starting point for each analysis is the multidimensional event log (MEL; step ①). It is a specific DWH which stores all the available event data in a cube-like data structure. Section 5 introduces its logical data model while Sect. 6 presents its relational-based realization.

By using OLAP operators (step ②), it is possible to filter (slice or dice) the MEL or to change its level of aggregation (roll-up or drill-down). This allows for the creation of flexible views on the event data. The query results in a set of cells where each cell contains a sublog. Each sublog is mined separately (step ③) using an arbitrary process discovery algorithm to create an individual process model reflecting the behavior of the sublog. Additionally, it is possible to enhance

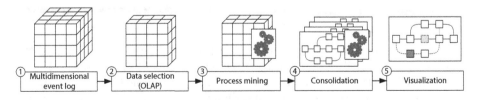

Fig. 4. Basic concept of PMCube

the process model with additional perspectives or to measure its quality using conformance checking techniques.

The OLAP query may result in multiple cells, leading to an unmanageable amount of process models. Therefore, PMCube introduces an optional step of consolidation (step ④), which aims to reduce the complexity of results. Its main idea is to automatically preselect a subset of potentially interesting process models by a heuristic. For example, assuming that big differences are more relevant to the analyst than minor differences between the models, it is possible to cluster similar process models and select a representative for each cluster. Alternatively, the process models can be selected by simply filtering them by their properties (e.g., occurrence of particular nodes). Finally, the process models are visualized (step ⑤). As MPM strongly benefits from comparing the different process models, it is not sufficient to visualize each model on its own. Therefore, PMCube provides several visualization techniques, e.g. merging two models into a difference model highlighting the differences between them. The concept of PMCube is presented in [18] in more detail.

5   Logical Data Warehouse Model

In contrast to the Process Cubes approach, the MEL maps the structure of event logs onto a data cube and organizes cases and events on different levels. Furthermore, the cells of the MEL do not contain sets of events, but a set of cases. The attributes of the cases are considered as dimensions forming the multidimensional data cube. Each combination of dimension values identifies a cell of the cube. According to the values of its attributes, each case is uniquely mapped onto a cell. However, some attributes represent highly individual features, e.g., the name of the patient. Mapping them onto dimensions results in sparse data cubes and does not add any benefit to the multidimensional analysis. On the contrary, these attributes might give valuable insights if the process model behavior is related to individual cases. Therefore, these non-dimensional attributes are directly attached to the case as so-called simple attributes.

Related to their respective case, the events are stored inside the MEL as well. Similar to the case attributes, the event attributes can be interpreted as dimensions, too. To avoid the aggregation of events from different, independent cases, the event attributes form an individual multidimensional data space for each case which is contained in the cell of the case. Figure 5 shows the relationship of these nested cubes. Each cell of the data cubes on the event level consists of a set of events identified by a combination of event dimension values. Similar to cases, non-dimensional event attributes are directly attached to the event as simple attributes. All dimensions, both on the case and event level, may have an arbitrary number of hierarchically structured dimension levels.

The OLAP queries like slice, dice, roll-up and drill-down are defined on a set of base operators like filtering (selection) and aggregation. Due to different semantics, the definition of the operators might vary between case and event level. Figure 6 illustrates this using the example of the aggregation operator.

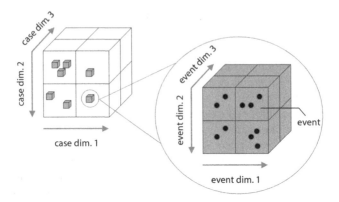

Fig. 5. Nested cubes

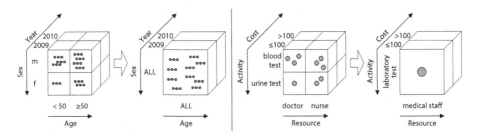

Fig. 6. Aggregation of cases (left) and events (right)

Aggregating cells on the case level creates the union of all the cells cases. For example, aggregating the cube on the left-hand of Fig. 6 along the dimensions *sex* and *age* results in a single cell containing all cases for both women and men of all age for a specific year. On the contrary, aggregating cells on the event level merges all events into a single, more abstract event. This is demonstrated on the right-hand side of Fig. 6, showing the aggregation of events along the dimensions *activity* and *resource*. Previously, various events are spread across different cells, each representing different kinds of activities performed by either doctors or nurses. The aggregation abstracts from the original events and replaces them by a single merged event. This style of aggregation can be useful if the analyst is only interested if a laboratory test was performed or not, regardless of which kind of test or how many tests were performed. Reducing the number of events may simplify the mined process model by reducing its number of nodes.

The MEL can be filtered by selection operators on both the case and the event level. On the event level, the selection predicate contains only event attributes (dimensions as well as simple attributes). This enables the analyst, for example, to remove all events representing non-medical activities to focus on the medical treatment process. On the case level, the MEL can be filtered by both case and event attributes. However, a quantifier ($\exists$ or $\forall$) must be specified for each event attribute of the selection predicate in order to specify whether the condition must hold for at least one event or for all events of a case. Alternatively, an aggregation function (*min*, *max*, *avg*, *sum*, or *count*) can be specified, e.g. to select only cases exceeding a maximum cost limit.

The MEL typically contains all available case and event attributes. However, most process mining techniques (i.e. process discovery algorithms) only need a small subset of attributes. To reduce the amount of data loaded from the MEL, the projection operator can be used to remove unneeded attributes. This may significantly speed up the OLAP query in case of slow database connections. In contrast to Process Cubes, the data model of our approach is a little more restrictive, e.g. it is not possible to change the case id during the analysis. However, it allows for a wider range of operations (e.g., selecting full cases based on event attributes) and a clear mapping onto the relational data model which is discussed in the following section.

6   Relational Data Warehouse Model

In contrast to traditional data cubes, the cells of the MEL do not contain single values but complex data. As available data warehousing tools are not capable of handling such data, MPM requires specific solutions storing event log data. The cells of the MEL consist of an arbitrary number of cases and events, which contradicts the MOLAP approach, where each cell typically represents a data point of fixed size. In contrast, ROLAP approaches allow for a more flexible modeling of complex data. Additionally, a ROLAP-based approach for MPM can benefit from various optimization techniques implemented in state-of-the-art DBMS. Therefore, we choose a ROLAP approach to realize the MEL.

Even though star schemes usually provide a better performance, we extend the traditional snowflake schema (cf. Sect. 2) to avoid redundancy which may lead to data anomalies. Figure 7 shows the generic database schema as an entity-relationship model. Similar to the traditional snowflake schema, there is a *fact*

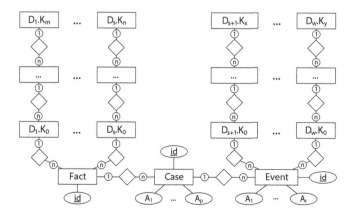

Fig. 7. Generic database schema of the MEL

table for storing the cells of the data cube. Each cell is identified by a unique combination of foreign keys referencing the cells dimension values. These values are stored in normalized dimension tables (e.g., $D_1.K_0$ to $D_1.K_m$ for a dimension D_1) to avoid redundancy. In contrast to the traditional snowflake schema, the fact table does not directly store the cells value, but a unique *id*. The data content of the cells, namely the cases, is normalized and stored in the *case* table, which also stores the simple attributes $(A_1$ to $A_p)$ of the cases. The corresponding cell of a case is referenced by the *fact id*. The events are normalized in an analogous manner and stored in the *event* table, which also holds the simple attributes of the event. Events can also have dimension attributes, which are stored in dimension tables similar to the case dimensions. However, the event table directly references the dimension tables, as dimension values might differ for events of the same case.

Figure 8 shows a possible database schema for the running example as an entity-relationship model. All case attributes are modeled as dimensions. While the sex and hospital dimension only consist of a single dimension level, the age and the year of treatment dimensions have additional dimension levels. These artificial dimension levels enable the analysts to group the patients by 5-year and 10-year classes respectively. For the events, only the activity attribute is mapped onto a dimension with two dimension levels. In this example, the *Activity_type* dimension level is introduced to define several types of activities, e.g., admission and discharge activities can be considered as organizational activities. All other event attributes (cost, time-stamp, and resource) are attached to the Event table as simple attributes. Note that there are also other possibilities to model the database schema for the running example, e.g., one could map the resource onto a dimension as well.

Figure 9 illustrates how the event data is loaded from the MEL and processed in PMCube. The starting point is an OLAP query which is defined by a user, e.g.,

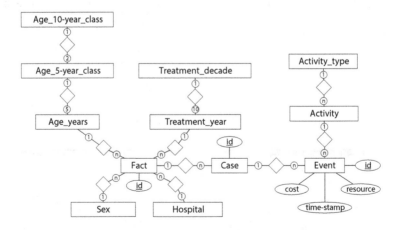

Fig. 8. Example database schema for the running example

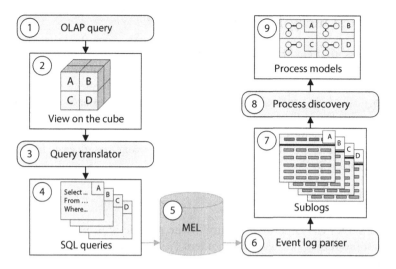

Fig. 9. Mapping an OLAP query to a set of SQL queries

through a graphical user interface (cf. ①). By this OLAP query, the user describes a logical view on the data cube (cf. ②). After that, the OLAP query is translated into a set of SQL queries (cf. ③ and ④). Each SQL query represents a cell defined by the OLAP query and expresses the appropriate filtering and aggregation operations. Section 7 presents the query translation in more detail. The SQL queries are sent to the MEL consisting of a relational database (cf. ⑤). The result of each query is a set of tuples, each tuple representing an event with all its (case and event) attributes. Immediately after the query result is sent back, the tuple set is parsed (cf. ⑥) and translated into a sublog (cf. ⑦) with the usual event log structure (cf. Fig. 1). Then the sublogs are mined using an arbitrary process discovery algorithm (cf. ⑧). To improve the performance, the sublogs are parsed and mined asynchronously. This means that the data is processed immediately after it has been loaded from the MEL. Finally, the process discovery results in a set of process models (cf. ⑨), one model for each cell.

Typically, the events in the traces of an event log are temporally ordered. This is mandatory to identify relationships between particular events. File-based event log formats like XES [4] usually imply this order by the position in the documents structure. However, relational databases store unordered multisets of tuples. To be able to restore the event order, PMCube requires the definition of an order-preserving attribute. By default, we use the event time-stamp for this. However, it might be possible that the time-stamp is missing or not precise enough to preserve the event order. Therefore, it is also possible to use other attributes, e.g., an explicit event index or the event id if this reflects the order of events.

7   Generic Query Patterns

PMCube aims to benefit from the various optimization techniques of state-of-the-art relational DBMS providing high performance and scalability. Therefore, PMCube expresses the filtering and aggregation operators within the SQL queries to push their processing to the database. PMCube uses a generic query pattern to map the cells defined by the OLAP query onto correspondent SQL queries. In Sect. 7.1, we first explain the general structure of the generic query pattern. After that, we show its extensions for event-based case filtering and event aggregation in Sects. 7.2–7.5.

7.1   General Structure

Listing 1.1 shows the general structure of the generic SQL pattern. To improve the understandability, we use placeholders ($<\dots>$) for particular parts of the pattern which will be explained in more detail.

```
1  SELECT <event log attributes>
2  FROM Fact
3      JOIN Case  ON Case.fact_id = Fact.id
4      JOIN Event ON Case.id = Event.case_id
5      <dimension joins>
6  WHERE <filter conditions>
7  ORDER BY Event.case_id, <sequence-preserving attribute>
```

Listing 1.1. Generic query pattern in SQL-like pseudo code

The placeholder *<event log attributes>* (line 1) is replaced by a list of all database attributes that should be loaded from the database. These database attributes can comprise values of dimension attributes and non-dimensional attributes, both on the case and event level. Representing the MEL's projection operator, it is possible to omit unneeded attributes by specifying a subset of attributes. This reduces the size of the data to be loaded from the database, leading to faster responses, especially if the data is transferred via a low bandwidth network connection. However, the case id, the sequence-preserving attribute, and the classifier (representing the name of the nodes in the process model) are mandatory and must be contained in the attribute list.

As the event data is spread across multiple database tables, it is necessary to join the tables to reconstruct the relationships between them. Therefore, the central tables (*fact*, *case*, and *event*) are joined (lines 2–4). Additionally, the fact table and the event table need to be joined with the dimension tables, to link the events with their respective dimension level values. The placeholder *<dimension joins>* (line 5) subsumes these join statements. Because join operations are very costly, we limit them to the tables that are required to filter the data or to retrieve the attributes specified in *<event log attributes>* (line 1). All other tables are omitted from *<dimension joins>* during query translation.

The placeholder <*filter conditions*> (line 6) subsumes all filtering operations, both on the case and the event level, as a conjunction of boolean expressions. Because each SQL query represents a particular cell, the dimensions forming the cube must be limited to their respective dimension values for this cell. For example, if a cell should represent all patients of the year 2015, <*filter conditions*> must contain an expression like DIM_TIME_YEAR.VALUE = 2015 (assuming that DIM_TIME_YEAR is the name of the table representing the time dimension at the year level and that VALUE is the name of an attribute of this table storing the values of the year). Finally, the tuples of the result table are sorted by the case id and the sequence-preserving attribute (line 7). This is done to restore the sequence of events for each case.

```
1   SELECT Event.case_id, Event.Timestamp, Activity.value,
2       Hospital.value, Sex.value, Treatment_year.value,
3           Event.resource, Event.cost
4   FROM Fact
5       JOIN Case ON Case.fact_id = Fact.id
6       JOIN Event ON Case.id = Event.case_id
7       JOIN Hospital ON Hospital.id = Fact.Hospital_id
8       JOIN Sex ON Sex.id = Fact.Sex_id
9       JOIN Year_of_treatment ON
10          Year_of_treatment.id = Fact.year_of_treatment_id
11              JOIN Activity ON Activity.id = Event.Activity_id
12  WHERE Hospital.value = 'Oldenburg'
13      AND Sex.value = 'female'
14      AND Year_of_treatment.value IN (2012, 2013, 2014)
15  ORDER BY Event.case_id, Event.timestamp
```

Listing 1.2. Example instantiation of the generic query pattern

Listing 1.2 shows an example instantiation of the generic query pattern based on the running example. Besides the case id, the time-stamp, and the event's activity, it also loads the name of the treating hospital, the sex of the patient, the year when the patient was treated, the resource performing the activity, and the individual costs related to an event (lines 1–3). In addition to the mandatory join of the fact, case and event tables (lines 4–6), this query joins the fact and event tables with the dimension tables (lines 7–11) in order to retrieve the selected attributes from the dimension tables. As the query is only defined for a single cell, the WHERE clause restricts the events to all female patients that were treated in the hospital in Oldenburg (lines 12–13). Additionally, the data is filtered by the year of the treatment, restricting the result to treatments in the years 2012 to 2014 (line 14). Finally, the events of the result set are ordered by their case and the time-stamp to reconstruct the traces of the event log (line 15).

Table 6 shows the query result when executing the query from Listing 1.2 on the event log L_1 (cf. Table 1). Due to the filter condition (only female patients who were treated in Oldenburg between 2012 and 2014), there are only three out

Table 6. Result of example query from Listing 1.2

Case id	Timestamp	Activity	Hospital	Sex	Year	Resource	Cost
1	01.02.2012 08:30	Admission	Oldenburg	Female	2012	Nurse	58
1	01.02.2012 09:42	Examination	Oldenburg	Female	2012	Doctor	161
1	01.02.2012 10:38	Blood test	Oldenburg	Female	2012	Nurse	142
1	01.02.2012 10:51	X-ray scan	Oldenburg	Female	2012	Doctor	61
1	02.02.2012 09:11	Surgery	Oldenburg	Female	2012	Doctor	66
1	03.02.2012 08:23	Surgery	Oldenburg	Female	2012	Doctor	50
1	04.02.2012 13:27	Discharge	Oldenburg	Female	2012	Nurse	137
3	01.02.2014 08:30	Admission	Oldenburg	Female	2014	Nurse	102
3	01.02.2014 09:42	Examination	Oldenburg	Female	2014	Doctor	163
3	01.02.2014 10:38	Blood test	Oldenburg	Female	2014	Nurse	168
3	01.02.2014 10:51	X-ray scan	Oldenburg	Female	2014	Doctor	168
3	02.02.2014 09:11	Surgery	Oldenburg	Female	2014	Doctor	140
3	04.02.2014 13:27	Discharge	Oldenburg	Female	2014	Nurse	98
11	01.02.2013 08:30	Admission	Oldenburg	Female	2013	Nurse	67
11	01.02.2013 09:42	Examination	Oldenburg	Female	2013	Doctor	41
11	01.02.2013 10:38	X-ray scan	Oldenburg	Female	2013	Doctor	178
11	01.02.2013 10:51	Urine test	Oldenburg	Female	2013	Nurse	176
11	02.02.2013 09:11	Surgery	Oldenburg	Female	2013	Doctor	105
11	03.02.2013 08:23	Surgery	Oldenburg	Female	2013	Doctor	40
11	04.02.2013 13:27	Discharge	Oldenburg	Female	2013	Nurse	157

of 20 cases selected. Each one of the resulting tuples represents a single event. The events are grouped by their related case and sequentially ordered by their case id and time-stamp. According to the projection (Listing 1.2, lines 1–2), the extracted data is restricted to a subset of the available attributes.

To filter cases by the attributes of their events, the *<filter conditions>* in Listing 1.1 (line 6) need to be extended by a subquery. The subquery selects the case ids of all cases meeting a particular condition. Due to the different kinds of case selection over event attributes ($\exists$, $\forall$, aggregation), there are differences in the patterns for the subqueries as well.

7.2   Case Filter Extension: $\exists$ Subquery

Listing 1.3 shows the subquery for the selection of cases with at least one event per case matching a condition. It simply selects all case ids of an event meeting the boolean condition given in line 3 (*<condition>*). Duplicates are removed using the UNIQUE key word, because more than one event of a case might match the condition. Usually, removing duplicates is a very expensive database operation. However, this overhead can be significantly reduced by defining an index on the case_id attribute in the database.

```
1   ... AND case_id IN (
2              SELECT UNIQUE case_id FROM Event
3              WHERE <condition>
4        ) ...
```

Listing 1.3. Subquery for selecting cases with at least one event matching a condition

Based on the running example, Listing 1.4 shows an example subquery that selects all cases for which at least one blood test has been performed. All other cases are omitted from the query result. For convenience, we use readable and meaningful string ids for the activity in this example.

```
1   ... AND case_id IN (
2              SELECT UNIQUE case_id FROM Event
3              WHERE Activity_id = 'blood test'
4        ) ...
```

Listing 1.4. Example subquery for selecting cases with at least one blood test event

Assuming that the WHERE clause in the example of the generic query pattern (Listing 1.2) is extended by the condition shown in Listing 1.4, the query will return a subset of the result from Table 6 which only comprises the tuples related to cases 1 and 3. The tuples for case 11 will be removed from the result, because no blood test was conducted for this patient.

7.3   Case Filter Extension: ∀ Subquery

If the condition must hold for each event of a case, the subquery shown in Listing 1.5 is used. Because SQL does not support such a selection, we use double negation. First, we select the ids of all cases that violate the condition expressed in <condition> (line 3). At this point, we also have to check all variables <v1> to <vn> used in <condition> for NULL values (lines 4–6). This is required because undefined attribute values are a violation of the condition as well which is however not covered by the condition in line 3. After we have selected the ids of all cases violating the condition, we only select that cases not contained in this subset (line 1).

```
1   ... AND case_id NOT IN (
2              SELECT UNIQUE case_id FROM Event
3              WHERE NOT <condition>
4                    OR <v1> IS NULL
5                    OR ...
6                    OR <vn> IS NULL
7        ) ...
```

Listing 1.5. Subquery for selecting cases where each event of a case matches a condition

Listing 1.6 shows an example subquery for selecting cases whose events were all executed by a nurse. As described above, the subquery selects all ids of all cases that have an event violating this condition, e.g., if an activity was performed by a doctor. Note that the attribute `resource` has to be checked for NULL values as events with an unknown resource violate the condition, too.

```
1   ... AND case_id NOT IN (
2           SELECT UNIQUE case_id FROM Event
3           WHERE NOT resource = 'nurse'
4               OR resource IS NULL
5   ) ...
```

Listing 1.6. Example subquery for selecting cases whose events were all executed by a nurse

Adding the subquery in Listing 1.6 to the WHERE clause of the generic query pattern example (Listing 1.2), will return an empty result set when executed on the example data from event log L_1. This is obvious as every patient already selected by the conditions defined in Listing 1.2 (female, treated between 2012 and 2014 in Oldenburg) has multiple events that where conducted by a doctor. Consequently, there is no case having all its events exclusively executed by nurses.

7.4   Case Filter Extension: Aggregation Subquery

Furthermore, PMCube allows for the selection of cases by aggregated event attributes. Assuming each event has an attribute representing the individual cost for the execution of its activity, it is possible to select all cases that, e.g., have at least an average cost per event of $100. This allows the analyst to define a very specific filtering of cases. Listing 1.7 shows the subquery to express this kind of filtering. The subquery groups all events by the id of their cases (line 3). After that, the condition is evaluated for each case. The placeholder <*condition*> (line 4) consists of a boolean expression which specify an aggregation over the grouped events. It is possible to use arbitrary SQL aggregation functions like SUM, AVG, MIN, or MAX for any event attribute.

```
1   ... AND case_id IN (
2               SELECT UNIQUE case_id FROM Event
3               GROUP BY case_id
4               HAVING <condition>
5           ) ...
```

Listing 1.7. Subquery for selecting cases using aggregations over event attributes

An example for selecting cases by aggregated event attribute values is given in Listing 1.8. It selects all cases whose events in average exceed the cost limit of $100 per event.

```
1   ... AND case_id IN (
2           SELECT UNIQUE case_id FROM Event
3           GROUP BY case_id
4           HAVING AVG (cost) > 100
5   ) ...
```

Listing 1.8. Example subquery for selecting cases whose events have an average cost of more than $100

Extending the WHERE clause of the example query in Listing 1.2 by the subquery from Listing 1.8 and executing it on event log $L1$ will return a subset of Table 6. The query result will only contain the events for the cases 3 and 11 as their average costs ($139.8 and $109.1, respectively) exceed the threshold of $100. The events related to case 1 will be removed because the average event costs of 96.4$ do not meet the condition.

7.5   Event Aggregation Extension

Finally, it is also possible to realize the aggregation of events (cf. Sect. 5) within the SQL query. For this operation, we extend the generic query pattern from Listing 1.1 at several points. First, we insert a GROUP BY $< attributes >$ statement between lines 6 and 7 to group the attributes that should be merged into a single high-level attribute. To avoid mixing events from different cases, the attribute list $<attributes>$ starts with the case id. Additionally, the list comprises all dimension attributes at their respective dimension level that should be targeted by the aggregation. Note that omitting a particular dimension from $<attributes>$ rolls up the data cube to the artificial root node ALL which describes the highest level of abstraction comprising all values of a dimension. E.g., inserting the statement GROUP BY Event.case_id, Activity.value aggregates all events of a case that represent the same activity to a new high-level activity.

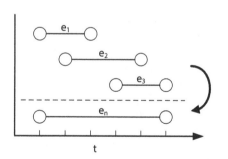

Fig. 10. Aggregating start and end time-stamps of events

The aggregated events have the same structure as the original events. Therefore, the event attributes of the original attributes have to be aggregated into a single value for each aggregated event. The aggregation of the dimension attributes is given by the structure of the dimension hierarchies. For each non-dimensional attribute, we individually select a SQL aggregation function depending on the semantics of the attribute. E.g., for the attribute *cost of activity* it makes sense to sum up the individual costs of the original events. This way, the new value will reflect the total cost of all events that are comprised by the aggregated event. However, depending on the analysis question, also other aggregation functions (e.g., average) might be meaningful, so it is a partially subjective choice.

Figure 10 illustrates the merging of the start and end time-stamps of events as another example. We use the MIN function for the start time-stamp and the MAX function for the end time-stamp. Consequently, the aggregated event e_n covers the same period of time like the single events e_1, e_2, and e_3. However, there might be some event attributes that cannot be aggregated in a meaningful way. To ensure that these attributes do not impede the aggregation, we propose to use the MIN function to aggregate the attributes even though these attributes will probably not contribute to the interpretation of results anymore. Additionally, all case attributes in the attribute list <*event log attributes*> (cf. Listing 1.1, line 1) have to be surrounded by aggregation function. This is due to the fact that SQL only allows for aggregations and grouping attributes after the SELECT statement if a grouping is used. We propose to use the MIN function for them, because all case attributes have the same value for each event of a case and the MIN function will preserve this value.

```
1   SELECT Event.case_id, min(Event.Timestamp) as alias_ts,
2           Activity.value, min(Hospital.value),
3           min(Sex.value), min(Treatment_year.value),
4           min(Event.resource), SUM(Event.cost)
5   FROM Fact
6       JOIN Case ON Case.fact_id = Fact.id
7       JOIN Event ON Case.id = Event.case_id
8       JOIN Hospital ON Hospital.id = Fact.Hospital_id
9       JOIN Sex ON Sex.id = Fact.Sex_id
10      JOIN Year_of_treatment ON
11          Year_of_treatment.id =Fact.year_of_treatment_id
12          JOIN Activity ON Activity.id = Event.Activity_id
13  WHERE Hospital.value = 'Oldenburg'
14      AND Sex.value = 'female'
15      AND Year_of_treatment.value IN (2012, 2013, 2014)
16  GROUP BY Event.case_id, Activity.value
17  ORDER BY Event.case_id, alias_ts
```

Listing 1.9. Extended example query aggregating events by their activity

Listing 1.9 shows an extended query based on the example query from Listing 1.2 that aggregates the events by the activity. This results in a sublog where each activity occurs at most once per trace, e.g., each trace has at most one *blood test* activity. In this example, we use the MIN function to aggregate the values of the hospital, the sex, the year of treatment dimensions because SQL only allows to select aggregated values or grouping attributes. As all case attributes have the same value for all events of a case, selecting the minimum does not change the results here. For the event attribute, we use the MIN function as well while we use the SUM function to calculate for the event costs. In contrast to the example presented in Fig. 10, we only have a single time-stamp related to an

Table 7. Result of example query from Listing 1.9 (event aggregation)

Case id	Timestamp	Activity	Hospital	Sex	Year	Resource	Cost
1	01.02.2012 08:30	Admission	Oldenburg	Female	2012	Nurse	58
1	01.02.2012 09:42	Examination	Oldenburg	Female	2012	Doctor	161
1	01.02.2012 10:38	Blood test	Oldenburg	Female	2012	Nurse	142
1	01.02.2012 10:51	X-ray scan	Oldenburg	Female	2012	Doctor	61
1	**02.02.2012 09:11**	**Surgery**	**Oldenburg**	**female**	**2012**	**doctor**	**116**
1	04.02.2012 13:27	Discharge	Oldenburg	Female	2012	Nurse	137
3	01.02.2014 08:30	Admission	Oldenburg	Female	2014	Nurse	102
3	01.02.2014 09:42	Examination	Oldenburg	Female	2014	Doctor	163
3	01.02.2014 10:38	Blood test	Oldenburg	Female	2014	Nurse	168
3	01.02.2014 10:51	X-ray scan	Oldenburg	Female	2014	Doctor	168
3	02.02.2014 09:11	Surgery	Oldenburg	Female	2014	Doctor	140
3	04.02.2014 13:27	Discharge	Oldenburg	Female	2014	Nurse	98
11	01.02.2013 08:30	Admission	Oldenburg	Female	2013	Nurse	67
11	01.02.2013 09:42	Examination	Oldenburg	Female	2013	Doctor	41
11	01.02.2013 10:38	X-ray scan	Oldenburg	Female	2013	Doctor	178
11	01.02.2013 10:51	Urine test	Oldenburg	Female	2013	Nurse	176
11	**02.02.2013 09:11**	**Surgery**	**Oldenburg**	**female**	**2013**	**doctor**	**145**
11	04.02.2013 13:27	Discharge	Oldenburg	Female	2013	Nurse	157

event. Therefore, we map the time-stamps of the aggregated events onto its
earliest occurrence using the minimum. However, this is a subjective choice.
Alternatively it is also possible to use other aggregation functions, e.g., the
maximum to map the events onto the latest occurrence. Note that the case
id and the activity dimension do not require an aggregation function because
both are used in the GROUP BY clause (line 16). Furthermore, the aggregated
time-stamp is renamed to alias_ts in order to use the aggregated value for
ordering the events.

Table 7 shows the result for executing the query from Listing 1.9 on the
example data of Event Log L_1. Its content is quite similar to the results shown
in Table 6. The only difference in this example is that the surgery events for
cases 1 and 11 are aggregated into a single event for each case (highlighted in
Table 7). For the merged event, the time-stamp of the first event is used. For the
cost attribute, the individual costs of the original events are summed up.

The case study reported in [18] revealed that the aggregation of events is
able to improve the fitness of the resulting process models if the frequency of
an activity per case does not correlate with its importance. For example, each
parameter of a blood test event may be represented by an individual event.
Algorithms which consider the ratio between the events, tend to overestimate
the importance of these events. The Fuzzy Miner [5], for example, will cluster
the less frequent events into one node while representing the more frequent but
less important blood test events as individual nodes. Aggregating the events

by their activity removes this bias because it restricts the maximum frequency per trace of each activity to one. On the contrary, the process model can also be significantly biased by the event aggregation itself. Especially if other events are located between the aggregated events in the trace, the dependencies to the aggregated events are lost. This may be indicated by a significant loss of fitness of the resulting process model. Therefore, the analysts are recommended to be aware of the possible bias of the process model and apply the event aggregation carefully.

8   Evaluation

We implemented our approach in a prototype called PMCube Explorer [17]. Due to a generic, plug-in-based interface, arbitrary DBMS can be used to store the data. We conducted a case study and a number of experiments to evaluate our approach. We reported the results of the case study in [18]. In this section, we focus on the experiments measuring the run-time performance of our approach.

For our experiments, we used the data set of the case study. It consists of 16,280 cases and 388,395 events and describes a process in a large German hospital of maximum care. For a more detailed description of the data, we refer to Vogelgesang and Appelrath [18]. We created multiple subsets of that data, each consisting of a particular number of events. While creating these subsets, we kept the cases as a whole in order to avoid splitting up the traces into fragments. Table 8 shows the number of cases and events for each data set. Figure 11 clearly shows the linear relationship between the number of cases and the number of events of the data sets.

To evaluate PMCube, we also performed similar tests with the PMC tool[1] as the state-of-the-art implementation of the Process Cubes approach. Event Cubes were not considered in the experiments because that approach is very different (e.g., no creation of sublogs) which makes it incomparable to PMCube and Process Cubes. All experiments were conducted on a laptop with Intel Core i5-2520M 2,5 GHz CPU, 16 GB DDR3 RAM, and ATA SSD running on 64-bit Windows 7. For the MEL, we used a local installation of Microsoft SQL Server 2012. We used the following OLAP queries derived from our case study.

Q1: We used the dimensions *medical department* and *urgency* to create the cube and removed all cases with the value *unknown* for the *urgency* dimension. This query results in a cube with 12 cells.

Table 8. Number of cases and events of the evaluation's data sets

Sample	1	2	3	4	5	6	7	8
Events	50,000	100,000	150,000	200,000	250,000	300,000	350,000	388,395
Cases	2,154	4,332	6,340	8,448	10,418	12,538	14,734	16,280

---

[1] http://www.win.tue.nl/~abolt/userfiles/downloads/PMC/PMC.zip, downloaded on June, 16th 2015.

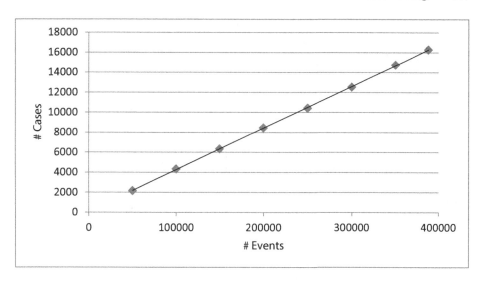

Fig. 11. Linear relationship of the number of cases and events for the evaluation's subsets of data

Q2: The dimensions *medical department* and *reason for discharge* form the cube. No filtering was used. Depending on the used subset of the data, this query creates 40 to 56 cells for PMC which only considers dimension values reflected in the data set. In contrast, PMCube Explorer considers each possible dimension value which is defined in the database, whether it is reflected by the events or not. Therefore, it procudes 92 cells for each subset.

Q3: The dimensions *age* (in 10-year-classes) and *sex* form the cube. All cases with values *unknown* and *older than 100 years* were removed. This query results in 30 cells.

Q4: We used the dimensions *urgency* and *type of admittance* (in-patient or not) to create the cube, filtering the value *unknown* for both dimensions. The resulting data cube for this query contains six cells.

Both tools, PMC and PMCube Explorer, vary in the set of supported operations. For example, PMCube Explorer provides the aggregation of events while PMC does not support this. Consequently, we only used queries that could be defined and answered by both tools. All queries partition the data by a number of case attributes into sublogs and filter the data by neglecting particular cells. As the definition of these operations is similar for both approaches, the resulting sublogs are congruent. However, PMCube Explorer may return more cells than PMC (cf. query Q2) as it also considers predefined dimension values that are not represented in the data. However, these additional cells do not affect the results of other cells and only require minimal overhead because they only contain empty sublogs.

Reflecting the overall waiting time for the analysts, we measured the total run-time for the processing of the OLAP query, discovering the model using

Table 9. Average run-times in seconds

Query	Events (in thousand)	50	100	150	200	250	300	350	388.395
Q1	PMC (min)	28.7	98.6	212.9	356.2	560.6	792.8	1013.0	1263.3
	PMCube Explorer (seq)	7.1	11.0	15.5	20.5	25.6	30.4	35.5	38.8
	PMCube Explorer (async)	5.0	7.3	10.2	13.5	17.1	20.3	24.5	26.4
Q2	PMC (min)	31.7	108.9	222.3	387.1	561.2	800.2	1087.2	1319.9
	PMCube Explorer (seq)	8.7	13.3	19.6	26.1	30.5	36.1	40.7	45.0
	PMCube Explorer (async)	7.3	10.0	14.6	18.5	21.1	24.6	27.3	29.8
Q3	PMC (min)	31.0	101.8	214.1	363.1	549.8	761.4	1043.4	1302.7
	PMCube Explorer (seq)	8.3	12.1	17.0	21.8	26.9	31.7	36.8	40.1
	PMCube Explorer (async)	6.5	8.9	11.4	14.1	17.3	20.1	22.4	25.3
Q4	PMC (min)	28.0	96.9	203.5	350.1	534.2	755.5	1021.0	1269.3
	PMCube Explorer (seq)	4.7	8.7	13.5	18.9	24.7	30.1	35.9	41.6
	PMCube Explorer (async)	3.7	6.9	10.7	15.4	20.2	25.4	30.0	35.0

Inductive Miner [6], and visualizing the process models of all cells in a matrix using process trees. Because some preliminary test runs with small data sets indicated unacceptable run-times of several hours for PMC with bigger data sets, we only used the minimum set of dimensions for PMC to improve its performance. This means, that we omitted all dimensions from the data cube which were not used to express the OLAP query. However, for our approach PMCube, we kept all available dimensions in the data cube. Even though this might overrate the performance of PMC, the affect on run-time should be insignificant because dimension tables are not joined unless they are used in the OLAP query.

Table 9 shows the average run-time in seconds of ten runs for both tools and each combination of query and data set. To evaluate the performance improvement of the asynchronous process discovery, we also performed the experiments with an alternative configuration of PMCube Explorer with a sequential processing of cells, i.e., the processing of a cell is completed before the next cell is loaded from the database. Note that the last column of Table 9 shows the measured values for the full data set.

The values in Table 9 show that the measured run-times of PMCube Explorer, both for asynchronous as well as sequential processing of cells, are significantly shorter than the run-times of PMC. E.g., PMCube Explorer needs between 25 and 45 s to process the queries for the complete data set while PMC requires more than 21 min for it.

Figure 12 shows the run-times (in seconds) of the queries Q1-Q4 over the number of events. It reveals a polynomial incline for the PMC tool and a linear incline for the PMCube Explorer with asynchronous process discovery. Comparing the four charts to each other, PMC as well as PMCube Explorer show a similar run-time behavior for all queries. Figure 13 compares the run-time of both configurations of PMCube Explorer for the queries Q1-Q4. It confirms that the run-time increases linearly by the number of events. Additionally, it clearly shows the performance benefit of the asynchronous processing of events which

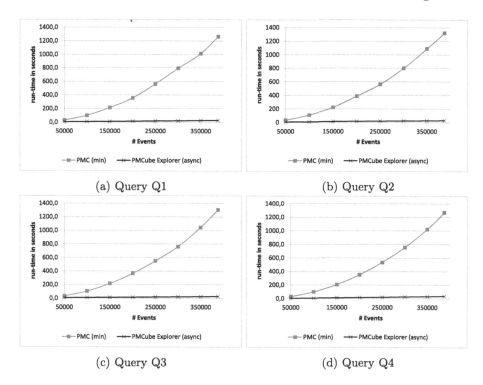

(a) Query Q1

(b) Query Q2

(c) Query Q3

(d) Query Q4

Fig. 12. Comparing average run-time of PMCube Explorer (async) with PMC (min) for queries Q1-Q4

increases by the number of events as well, even though the advantage of the asynchronous processing varies for each query.

Table 10 shows the measured loading times in seconds for PMC and both configurations of PMCube Explorer. The measured values reflect the time needed for processing the query and returning a set of sublogs. Comparing the different approaches clearly shows that the loading times of PMCube Explorer are significantly lower than the loading times of PMC. However, the loading times for PMCube Explorer (seq) are shorter than for PMCube Explorer (async). At first glance, this is in contrast to the overall processing times shown in Table 9, which are shorter for the asynchronous processing. These differences can be explained by the additional overhead required for the asynchronous processing. In the evaluation setting, all processes – also the database management system – are running on the same machine in parallel. Consequently, the loading of data (including the parsing of SQL results), the discovery of processes (implemented as an individual process for each cell) and the data processing of the database are competing for the same resources, especially CPU. Therefore, the loading process of PMCube Explorer may be waiting for resources while previously started process discovery threads are occupying the CPU cores. Due to this waiting times and the additional effort for process switches, the loading times are

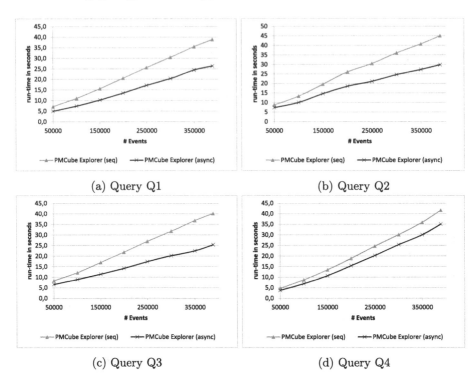

(a) Query Q1   (b) Query Q2

(c) Query Q3   (d) Query Q4

Fig. 13. Comparing average run-time of PMCube Explorer (async) with PMCube Explorer (seq) for queries Q1-Q4

Table 10. Average loading times in seconds

Query	Events (in thousand)	50	100	150	200	250	300	350	388.395
Q1	PMC (min)	27.9	97.4	211.6	354.8	558.7	790.8	1010.6	1260.7
	PMCube Explorer (seq)	1.9	3.4	4.9	6.6	8.0	9.7	11.5	12.7
	PMCube Explorer (async)	2.5	4.6	6.7	9.0	11.0	13.1	15.2	17.8
Q2	PMC (min)	30.5	107.1	220.4	385.0	558.4	797.4	1083.8	1316.4
	PMCube Explorer (seq)	2.9	4.7	6.6	8.8	10.7	12.8	15.2	16.6
	PMCube Explorer (async)	4.0	6.3	9.7	12.6	16.0	19.1	21.8	24.2
Q3	PMC (min)	29.8	100.2	212.5	361.5	547.4	759.1	1040.5	1299.8
	PMCube Explorer (seq)	2.4	4.1	5.8	7.7	9.4	11.2	13.9	14.5
	PMCube Explorer (async)	3.1	5.6	8.0	10.6	13.2	15.9	18.4	21.2
Q4	PMC (min)	27.4	96.0	202.4	348.8	532.6	753.7	1018.8	1267.1
	PMCube Explorer (seq)	1.6	3.0	4.4	5.9	7.2	8.5	10.2	11.4
	PMCube Explorer (async)	1.7	3.2	4.7	6.3	7.9	9.4	11.0	12.1

longer in case of asynchronous processing. However, as they significantly reduce the overall run-time for processing a query, these delays while loading the data are justified.

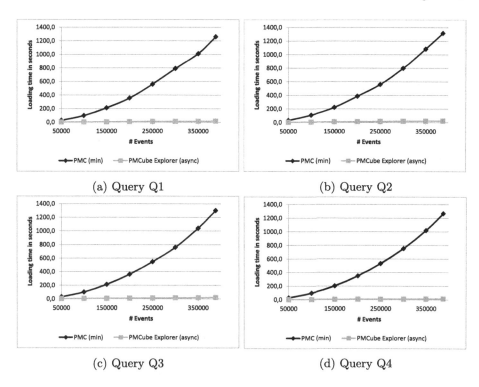

(a) Query Q1   (b) Query Q2

(c) Query Q3   (d) Query Q4

Fig. 14. Comparing average loading time of PMCube Explorer (async) with PMC (min) for queries Q1–Q4

Figure 14 shows the average loading times in seconds over the number of events for PMC and PMCube Explorer (async) for queries Q1–Q4. The charts reveal the same behavior as the overall run-time depicted in Fig. 12. While the loading time grows polynomially with the number of events for PMC, it grows linearly for PMCube Explorer (async). This clearly shows that the PMCube Explorer's advantage in run-time can be traced back to the data storage and management based on the relational data warehouse. The charts presented in Fig. 15 show the average loading times in seconds over the number of events for both configurations of PMCube Explorer for queries Q1–Q4. They also confirm the linear incline of the loading time by the number of events. Furthermore, the charts indicate that the measured loading times of PMCube Exporer (async) are bigger than the loading times of PMCube Explorer (seq). However, comparing the charts also shows that the difference varies between the queries. While the difference between both configurations is significant for queries Q1–Q3, there is only a slight difference for Q4.

Due to the clear linear relationship between the number of events and the number cases of the evaluation's data sets (cf. Table 8 and Fig. 11), the observed run-time behavior can also be related to the number cases. Consequently, the

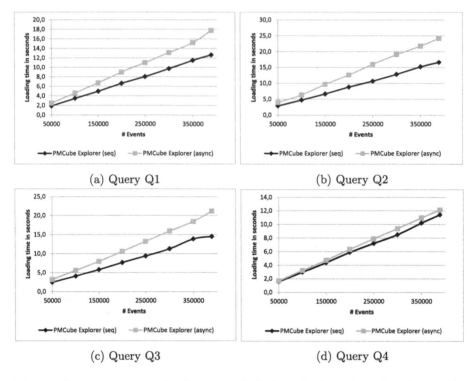

(a) Query Q1   (b) Query Q2

(c) Query Q3   (d) Query Q4

Fig. 15. Comparing average loading time of PMCube Explorer (async) with PMCube Explorer (seq) for queries Q1–Q4

run-times increases polynomially by the number of cases for PMC, while it increases linearly for PMCube Explorer.

9 Conclusions

Multidimensional process mining adopts the concept of data cubes to the field of process mining. Even though it is not time-critical, performance is a vital aspect due to its explorative characteristics. In this paper, we presented the realization of our approach PMCube using a relational DBMS. The logical data model is mapped onto a generic relational database schema. We use generic query patterns to express the OLAP queries by a separated SQL query for each cell. The experiments reported in this paper show, that PMCube provides a significantly better performance than PMC, the state-of-the-art implementation of the Process Cubes approach. Additionally, the performance of our approach seems to scale linearly by the number of events, promising acceptable processing times with bigger amounts of data. Nevertheless, further improvements of performance might be possible, e.g., by denormalizing the relational schema (similar to a star schema), which should be evaluated by future research. Furthermore, it should be investigated how to improve the flexibility of our approach, e.g., how to reflect multiple case ids in the database schema without loosing performance.

References

1. Bolt, A., van der Aalst, W.M.P.: Multidimensional process mining using process cubes. In: Gaaloul, K., Schmidt, R., Nurcan, S., Guerreiro, S., Ma, Q. (eds.) CAISE 2015. LNBIP, vol. 214, pp. 102–116. Springer, Heidelberg (2015). doi:10.1007/978-3-319-19237-6_7

2. Ekanayake, C.C., Dumas, M., García-Bañuelos, L., Rosa, M.: Slice, mine and dice: complexity-aware automated discovery of business process models. In: Daniel, F., Wang, J., Weber, B. (eds.) BPM 2013. LNCS, vol. 8094, pp. 49–64. Springer, Heidelberg (2013). doi:10.1007/978-3-642-40176-3_6

3. Golfarelli, M., Rizzi, S.: Data Warehouse Design: Modern Principles and Methodologies, 1st edn. McGraw-Hill Inc., New York (2009)

4. Günther, C.W.: XES Standard Definition, March 2014. http://www.xes-standard.org/xesstandarddefinition

5. Günther, C.W., van der Aalst, W.M.P.: Fuzzy mining – adaptive process simplification based on multi-perspective metrics. In: Alonso, G., Dadam, P., Rosemann, M. (eds.) BPM 2007. LNCS, vol. 4714, pp. 328–343. Springer, Heidelberg (2007). doi:10.1007/978-3-540-75183-0_24

6. Leemans, S.J.J., Fahland, D., van der Aalst, W.M.P.: Discovering block-structured process models from event logs - a constructive approach. In: Colom, J.-M., Desel, J. (eds.) PETRI NETS 2013. LNCS, vol. 7927, pp. 311–329. Springer, Heidelberg (2013). doi:10.1007/978-3-642-38697-8_17

7. Maggi, F.M., Dumas, M., García-Bañuelos, L., Montali, M.: Discovering data-aware declarative process models from event logs. In: Daniel, F., Wang, J., Weber, B. (eds.) BPM 2013. LNCS, vol. 8094, pp. 81–96. Springer, Heidelberg (2013). doi:10.1007/978-3-642-40176-3_8

8. Neumuth, T., Mansmann, S., Scholl, M.H., Burgert, O.: Data warehousing technology for surgical workflow analysis. In: Proceedings of the 2008 21st IEEE International Symposium on Computer-Based Medical Systems, CBMS 2008, pp. 230–235. IEEE Computer Society, Washington, D.C. (2008)

9. Niedrite, L., Solodovnikova, D., Treimanis, M., Niedritis, A.: Goal-driven design of a data warehouse-based business process analysis system. In: Proceedings of the 6th Conference on 6th WSEAS International Conference on Artificial Intelligence, Knowledge Engineering and Data Bases, AIKED 2007, vol. 6. pp. 243–249. World Scientific and Engineering Academy and Society (WSEAS), Stevens Point (2007)

10. Nooijen, E.H.J., Dongen, B.F., Fahland, D.: Automatic discovery of data-centric and artifact-centric processes. In: Rosa, M., Soffer, P. (eds.) BPM 2012. LNBIP, vol. 132, pp. 316–327. Springer, Heidelberg (2013). doi:10.1007/978-3-642-36285-9_36

11. Ribeiro, J.T.S., Weijters, A.J.M.M.: Event cube: another perspective on business processes. In: Meersman, R., et al. (eds.) OTM 2011. LNCS, vol. 7044, pp. 274–283. Springer, Heidelberg (2011). doi:10.1007/978-3-642-25109-2_18

12. Schönig, S., Rogge-Solti, A., Cabanillas, C., Jablonski, S., Mendling, J.: Efficient and customisable declarative process mining with SQL. In: Nurcan, S., Soffer, P., Bajec, M., Eder, J. (eds.) CAiSE 2016. LNCS, vol. 9694, pp. 290–305. Springer, Heidelberg (2016). doi:10.1007/978-3-319-39696-5_18

13. van der Aalst, W.M.P.: Process Mining: Discovery, Conformance and Enhancement of Business Processes. Springer, Berlin (2011)
14. van der Aalst, W.M.P.: Process cubes: slicing, dicing, rolling up and drilling down event data for process mining. In: Song, M., Wynn, M.T., Liu, J. (eds.) AP-BPM 2013. LNBIP, vol. 159, pp. 1–22. Springer, Heidelberg (2013). doi:10.1007/978-3-319-02922-1_1
15. van der Aalst, W., et al.: Process mining manifesto. In: Daniel, F., Barkaoui, K., Dustdar, S. (eds.) BPM 2011. LNBIP, vol. 99, pp. 169–194. Springer, Heidelberg (2012). doi:10.1007/978-3-642-28108-2_19
16. van Dongen, B.F., Shabani, S.: Relational XES: data management for process mining. In: Proceedings of the CAiSE 2015 Forum at the 27th International Conference on Advanced Information Systems Engineering Co-located with 27th International Conference on Advanced Information Systems Engineering (CAiSE 2015), Stockholm, Sweden, 10 June 2015, pp. 169–176 (2015). http://ceur-ws.org/Vol-1367/paper-22.pdf
17. Vogelgesang, T., Appelrath, H.: Multidimensional process mining with PMCube explorer. In: Daniel, F., Zugal, S. (eds.) Proceedings of the BPM Demo Session 2015 Co-located with the 13th International Conference on Business Process Management (BPM 2015), Innsbruck, Austria, 2 September 2015, CEUR Workshop Proceedings, vol. 1418, pp. S.90–S.94. CEUR-WS.org (2015). http://ceur-ws.org/Vol-1418/paper19.pdf
18. Vogelgesang, T., Appelrath, H.-J.: PMCube: a data-warehouse-based approach for multidimensional process mining. In: Reichert, M., Reijers, H.A. (eds.) BPM 2015. LNBIP, vol. 256, pp. 167–178. Springer, Heidelberg (2016). doi:10.1007/978-3-319-42887-1_14

Author Index

Printed in the United States
By Bookmasters